Mit **„BestMasters“** zeichnet Springer die besten Masterarbeiten aus, die an renommierten Hochschulen in Deutschland, Österreich und der Schweiz entstanden sind. Die mit Höchstnote ausgezeichneten Arbeiten wurden durch Gutachter zur Veröffentlichung empfohlen und behandeln aktuelle Themen aus unterschiedlichen Fachgebieten der Naturwissenschaften, Psychologie, Technik und Wirtschaftswissenschaften. Die Reihe wendet sich an Praktiker und Wissenschaftler gleichermaßen und soll insbesondere auch Nachwuchswissenschaftlern Orientierung geben.

Springer awards **“BestMasters”** to the best master’s theses which have been completed at renowned Universities in Germany, Austria, and Switzerland. The studies received highest marks and were recommended for publication by supervisors. They address current issues from various fields of research in natural sciences, psychology, technology, and economics. The series addresses practitioners as well as scientists and, in particular, offers guidance for early stage researchers.

More information about this series at https://link.springer.com/bookseries/13198

Praveen Manoharan

Universal Relations for Binary Neutron Star Mergers with Long-lived Remnants

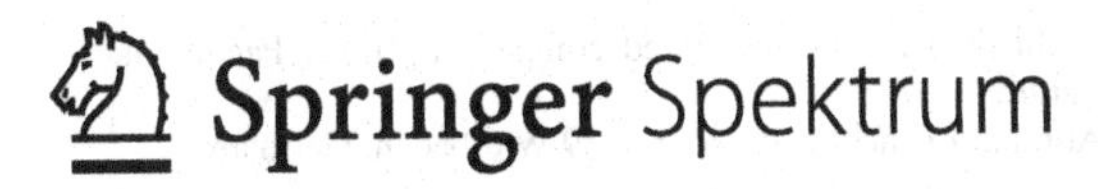

Praveen Manoharan
Tübingen, Germany

ISSN 2625-3577 ISSN 2625-3615 (electronic)
BestMasters
ISBN 978-3-658-36840-1 ISBN 978-3-658-36841-8 (eBook)
https://doi.org/10.1007/978-3-658-36841-8

Responsible Editor: Marija Kojic
This Springer Spektrum imprint is published by the registered company Springer Fachmedien Wiesbaden GmbH part of Springer Nature.
The registered company address is: Abraham-Lincoln-Str. 46, 65189 Wiesbaden, Germany

Abstract

In the last 25 years, an extensive body of work has developed various equation of state independent—or (*approximately*) *universal*—relations that allow for the inference of neutron star parameters from gravitational wave observations. These works, however, have mostly been focused on singular neutron stars, while our observational efforts at the present, and in the near future, will be focused on binary neutron star (BNS) mergers.

In light of these circumstances, the last five years have also given rise to more attempts at developing universal relations that relate BNS pre-merger neutron stars to stellar parameters of the post-merger object, mostly driven by numerical relativity simulations. Inspired by these works, we present, in this thesis, a first attempt at perturbatively deriving universal relations for binary neutron star mergers with long-lived neutron star remnants.

We succeed in confirming previous results relating pre-merger binary tidal deformabilities to the f-mode frequency of the post-merger object. Combining this result with recent advances of computing the f-mode frequency of fast rotating neutron stars, we also derive a combined relation that relates the pre-merger binary tidal deformability of a BNS to the effective compactness of a long-lived neutron star remnant. Finally, we also propose a direct relation between these quantities with improved accuracy.

The results presented in this thesis allow for the prediction of the stellar parameters of a long-lived remnant neutron star from the observation of gravitational waves emitted during the pre-merger phase.

Contents

Abbreviations & Notation

BNS	Binary Neutron Star (System)
c	Speed of Light in Vacuum, 2.9979×10^{10} cm s^{-1}
EM	Electromagnetic (Observations)
EoS	Equation of State
G	Newton's Gravitational Constant, 6.6743×10^{-8} cm^3 g^{-1} s^{-2}
GW	Gravitational Wave
$M_\odot$	Solar Mass, 1.989×10^{33} g
RMSE	Root-Mean-Squared Error (cf. Section 3.5)
C	Compactness, $C = \frac{M_0}{R}$
k_2	Quadrupolar Love Number
M	Total Gravitational Mass of BNS, $M = M_1 + M_2$, and of Long-lived Remnant
M_b	Baryon Mass of a Single Neutron Star
M_0	Gravitational Masses of Single Neutron Star
M_1, M_2	Gravitational Masses of Pre-merger Neutron Stars in BNS
q	Mass Ratio of BNS, $q = \frac{M_1}{M_2} \leq 1$
R	Radius of Single Neutron Star
ζ	Symmetric Mass Ratio of BNS, $\zeta = \frac{M_1 M_2}{(M_1+M_2)^2} = \frac{q}{(1+q)^2}$
Ω	Angular Rotation Rate of a Neutron Star
λ	Tidal Deformability, $\lambda = \frac{2}{3} k_2 R^5$
$\tilde{\Lambda}$	Binary Tidal Deformability (cf. Equation (3.33))
σ	f-mode Frequency
σ^s, σ^u	Stable (Co-rotating) and Unstable (Counter-rotating) f-mode Frequency
ω	Angular f-mode frequency, $\omega = 2\pi\sigma$

List of Figures

List of Tables

Introduction

1

The successful detection of gravitational waves (GWs) from binary neutron star (BNS) mergers through the LIGO-VIRGO detectors [AoLV17, AoLV20] has enabled a new avenue into probing and understanding the structure of neutron stars: through the observation of these mergers, we can put constraints on the bulk properties of neutron stars [RMW18, MWRS18, BBV+20], which in turn will allow us to uncover their true equation of state (EoS).

An important tool for this task are EoS independent—or *(approximately) universal*—relations for neutron stars: they relate quantities that can readily be extracted from the GW signal—and also potential electromagnetic (EM) observation—of BNS mergers to other bulk properties of neutron stars without requiring knowledge about the underlying EoS. Developing highly accurate universal relations for BNS mergers is therefore instrumental to further our understanding of neutron stars through GW (and EM) observations.

Such universal relations were first developed for the case of single neutron stars: starting with various polytropic EoS, Andersson and Kokkotas [AK98] proposed EoS independent relations between the primary oscillation mode frequencies, which determine the peaks in the spectrum of the gravitational waves emitted by an oscillating neutron star [KS92, AKK96], and the average density of a non-rotating neutron star, laying the foundation for the astroseismology of compact stars.

Their results were later followed up by rich line of work that e.g. examined the role of other oscillation modes [BBF99, BFG04], improved the accuracy of the universal relations [TL05b, LLL10], or extended these results to modern, realistic EoSs [CdK15] as well as rotating neutron stars [GK11, DGKK13, DK15, KK20b].

Another line of research centered around the tidal deformability of neutron stars, i.e. their response to the external tidal field induced by, e.g., their BNS companion: Yagi and Yunes [YY13] relate in their I-Love-Q relation, with high accuracy and universality, the tidal deformability of a non-rotating neutron star to its moment

P. Manoharan, *Universal Relations for Binary Neutron Star Mergers with Long-lived Remnants*, BestMasters, https://doi.org/10.1007/978-3-658-36841-8_1

of inertia and quadrupole moment. Due to the well-measurable effect of the tidal deformability on the phase evolution of the gravitational wave signal emitted by neutron stars [FH08], the I-Love-Q relation also spawned a significant amount of follow up work that tries to extended it to other EoSs [MOHN17, BWG+20], rotating neutron stars [DYSK14, CDGS14, PA14] and various other configurations [MCF+13, HCPR14].

Furthermore, since both, the oscillation mode frequencies and the tidal deformability take an important role in the universal relations for neutron stars—essentially characterizing the dynamical properties of oscillating neutron stars—Chan et al. [CSLL14] investigated and found universal relations directly between these two quantities, highlighting their close relationship.

Only in the last five years, enabled by the progress in numerical relativity simulations, have universal relations been developed for binary neutron stars mergers: these relations primarily connect the tidal deformability of the pre-merger stars—encoded, e.g., in the *binary tidal deformability* [Fav14]—to stellar parameters of the post-merger remnant [BDN15, RT16].

More recently, Kiuchi et al. [KKK+20] have shown that existing universal relations for BNS mergers suffer from systematic errors caused by, e.g., only considering even ratios between the masses of the pre-merger stars (a common assumption that had been called into question after the observation of `GW170817` [AoLV17]). Allowing for a wider range of mass ratios than in previous works, they propose alternative relations between the binary tidal deformability and post-merger remnant parameters with high accuracy (with maximum relative error at the order of 10^{-2}).

Vretinaris et al. [VSB20] recently also investigated empirical relations for BNS mergers based on the extensive coRE data set [DRB+18] of numerical relativity gravitational wave simulations. Covering a wide range of mass ratios, they find an extensive set of universal relations involving the various peak frequencies of the post-merger gravitational wave signal, involving, e.g., the chip mass and characteristic radius of a 1.6 solar mass neutron star. In particular, they also find universal relations between the binary tidal deformability and the primary f-mode frequency of the post-merger signal (as in [KKK+20]), however this time involving the chirp mass of the BNS.

Investigating a wider range of the BNS parameter space through numerical relativity simulations, however, is limited by its high computational cost. An alternative approach to deriving universal relations for BNS mergers is therefore highly desirable. In particular for the case where the BNS merger results in a long-lived neutron star remnant (instead of collapsing to a black Hole), perturbative calculations might allow for more efficient analyses.

1.1 Problem Statement

In this thesis, we evaluate an alternative approach to developing universal relations for BNS mergers with long-lived remnants. Instead of performing costly numerical relativity simulations to evaluate the full merger process, in particular the exact moment of the merger, we compare the results of independent perturbative calculations performed within the pre- and post-merger phases to obtain new universal relations.

To this end, we represent the BNS merger with a simplified model: two irrotational neutron stars with masses M_1 and M_2 merge, resulting in a long-lived neutron star remnant of total mass $M = M_1 + M_2$, represented by a cold, super-massive and uniformly rotating neutron star. Our model allows for a wide range of pre-merger mass ratios, post-merger remnant rotation rates, and EoSs. By separately computing and then comparing the tidal deformabilities of the pre-merger neutron stars and the oscillation modes and other stellar parameters of the remnant, we are able to analyze their relation and derive new universal relations.

We will ourselves perform the calculations for the pre-merger phase, where we compute the tidal deformabilities of the pre-merger stars using the formalism put forward by Hinderer [Hin08, Hin09]. For the treatment of the rapidly rotating, long-lived remnant, we rely on the extensive data set of rotating neutron star models underlying the work by Krüger and Kokkotas [KK20b], who, in particular, compute the f-mode frequency for fast-rotating neutron stars without approximation.

Using this method, we succeed in deriving a similar relation as proposed by Kiuchi et al. [KKK+20] between the binary tidal deformability $\tilde{\Lambda}$ of the BNS and the f-mode frequency of the remnant, however with different coefficients that depend on, both, the mass ratio q and the angular rotation rate Ω of the remnant neutron star. Combining this result with a universal relation between the f-mode frequency and the *effective compactness* η for fast-rotating neutron stars [LLL10, DK15, KK20b], we also derive a combined relation that relates $\tilde{\Lambda}$ to η.

Finally, by directly relating these quantities without going through the f-mode, we obtain a universal relation of the form

$$\log\left[M^5\eta\right] = a(q,\hat{\Omega})M^5\tilde{\Lambda}^{-\frac{1}{5}} + b(q,\hat{\Omega}) \tag{1.1}$$

with $\hat{\Omega} = M\Omega$. This relation achieves improved accuracy, reaching an average relative error of 1.7% across all EoSs we consider in this thesis.

While our approach allows for a wide range of parameters for the long-lived remnant, a number of works exists that put constraints on them: Radice et al. [RPBZ18], for instance, constrain the remnant's rotation period based on its baryon mass. We adopt this constraint and investigate potential improvements to the above relation. While our findings show that this constraint improves the accuracy of our relation only slightly, this extension does result in a relation with only one free parameter (the mass ratio), which improves the usability of our relation.

As a side-product of our calculations, we also investigate the relation between the tidal deformability and f-mode frequency for single, rotating neutron stars. Since the theory behind the tidal deformability of rotating neutron stars is still under investigation [PGMF18], we compare the non-rotating tidal deformability to the f-mode frequency of a rotating star, and find a linear universal relation between these quantities that depends on the rotation rate. While this relation has no real practical relevance, it does shed some further light on the relationship between these two quantities. Furthermore, this scenario also draws parallels to the BNS case, where we compare the tidal deformability of the non-rotating pre-merger stars to parameters of the rotating remnant.

The results laid out in this thesis present a first step towards perturbatively finding universal relations between the pre-merger neutron stars and the potential long-lived remnant of a BNS merger. Due to the modularity of our approach, it can easily be extended to include more involved models, such phase-transitions between the pre- and post-merger EoSs [BBB+19], and differential rotation for the long-lived remnant [BSB18].

1.2 Outline

We begin by discussing the history of universal relations for (binary) neutron stars in more detail by taking a closer look at related work in Chapter 2. We then discuss the theory underlying our work and the general methodology we follow for our analyses in Chapter 3 and Chapter 4, respectively. After presenting the results of our investigation in form of new universal relations in Chapter 5, we conclude this thesis and give a final outlook on fruitful directions for future work in Chapter 6.

Note that, throughout the thesis, we will assume, unless otherwise stated, geometrized units in which $G = c = 1$.

2 Related Work

The work presented in this thesis is primarily motivated by the rich body of work developed in the last 20 years on the topic of *universal* relations that relate properties of neutron stars in an EoS independent manner. In the following, we outline some of these works involving the oscillations modes and tidal deformability of single neutron stars, before we tackle more recently proposed relations for binary neutron star mergers.

2.1 Neutron Star Oscillation Modes

The frequencies of the various oscillation modes of a neutron stars depend directly on its structure and composition. In the context of gravitational wave observations, we are particularly interested in the fundamental mode (or f-mode) which can drive the emission of gravitational wave by oscillating neutron stars [AK98]. The f-mode frequency can also be extracted with high accuracy from the primary peak in the spectrum of the gravitational waves emitted by the remnant of a binary neutron star (BNS) merger [SBZJ11].

As such, relations that allow us to infer bulk parameters of neutron stars from a measurement of the f-mode frequency can be valuable tools for better understanding the structure of neutron stars.

2.1.1 Non-rotating Neutron Stars

Andersson and Kokkotas [AK98] first investigate the reliance of the f-, p- and w-mode eigenfrequencies of a non-rotating neutron star on its equation of state. To this end, they compute these eigenfrequencies for non-rotating neutron stars of twelve

P. Manoharan, *Universal Relations for Binary Neutron Star Mergers with Long-lived Remnants*, BestMasters, https://doi.org/10.1007/978-3-658-36841-8_2

different EoSs. Their findings show a nearly linear relation between the f-mode frequency σ and the square-root of the average density $\bar{\rho}^{\frac{1}{2}} = \sqrt{\frac{M_0}{R^3}}$ of the neutron star. Their linear fit yields

$$\sigma(\text{kHz}) \approx 0.78 + 1.635 \left(\frac{\bar{M}_0}{\bar{R}^3}\right)^{\frac{1}{2}} \tag{2.1}$$

where $\bar{M}_0$ and $\bar{R}$ are the normalized mass and radius

$$\bar{M}_0 = \frac{M_0}{M_\odot} \qquad \bar{R} = \frac{R}{10\text{km}}$$

As such, we have that $\sigma \sim \rho^{\frac{1}{2}}$. Similarly, they find for the damping time τ_f of the f-mode the relation

$$\frac{1}{\tau_f} \approx \frac{\bar{M}_0^3}{\bar{R}^4}\left[22.85 - 14.65\left(\frac{\bar{M}_0}{\bar{R}}\right)\right]$$

thus relating the normalized damping time $\left(\frac{\tau_f \bar{M}_0^3}{\bar{R}^4}\right)^{-1}$ with the compactness $\frac{\bar{M}_0}{\bar{R}}$ of the neutron star. Similar results were also shown for the p-mode (weak universality) and the w-mode (slightly better universality) eigenfrequencies and their damping times. With this work Andersson and Kokkotas thus provide first evidence for universal relations between oscillation frequencies and other parameters of neutron stars.

Benhar et al. [BBF99] extend this analysis to the axial w-mode eigenfrequencies and find similar empirical relations as above. Later, Benhar et al. [BFG04] perform the same analysis on a different set of EoSs from non-relativistic nuclear many-body theory and relativistic mean field theory. They obtain similar universal relations as above for the f-mode frequency and damping time, as well as the p-mode frequency, with slight differences in the resulting coefficients of the linear fits caused by the different set of EoSs used. Due to the longer damping time of the f-mode and p-mode oscillations, they also remark that these seem to be more suited for the task of GW astroseismology.

This is further supported by Kokkotas et al. [KAA01], who establish the potential ability of modern GW detectors to detect GWs sourced from f-mode and p-mode neutron star oscillations, and the possibility of using such signals to separate different neutron star EoSs.

Scaled Eigenfrequencies

Tsui and Leung [TL05a, TL05b] propose an alternative approach to finding universal relations for the oscillation modes of neutron stars by considering the Tolman VII model, a less accurate, but more general abstraction for any neutron star EoS. They find that by scaling the f-mode frequency with the gravitational mass M_0 of a neutron stars, a relation of the form

$$M_0\omega = a\left(\frac{M_0}{R}\right)^2 + b\left(\frac{M_0}{R}\right) + c \tag{2.2}$$

where a, b and c are constants determined by their best fit, and ω is the angular f-mode frequency $\omega = 2\pi\sigma$, is made possible. As this relation relates the scaled frequency $M_0\omega$ to the compactness $C = \frac{M_0}{R}$, we now have a relation markedly different than the ones discussed above.

By realizing that both, the universal relations by Tsui and Leung [TL05a], as well as the empirical relations found by Lattimer and Schutz [LS05] (cf. Section 2.3), depend on the compactness of the neutron star, Lau et al. [LLL10] seek to combine these relations to obtain new universal relation between the f-mode frequency, mass and moment of inertia of neutron stars. At the same time they also want to resolve the issue that for quark stars (i.e. compact, self-bound stars with stiff EoS), these previous relations generally do not remain universal.

To this end, they propose to utilize the effective compactness

$$\eta = \sqrt{\frac{M_0^3}{I}} \tag{2.3}$$

as the independent parameter which the scaled f-mode frequency $M_0\sigma$ is related to. Computing the scaled f-mode frequency and the effective compactness for a range of soft and stiff EoSs, they find the nearly perfect, quadratic fit of the form

$$M_0\sigma = a + b\eta + c\eta^2 \tag{2.4}$$

Tsui and Leung argue that their relation provides a much better fit across the different EoS than the previous universal relations, as evidenced by their smaller root-mean-square error [TL05a].

Chirenti et al. [CdK15] attempt to extend the universality of these relations by considering a larger set of modern EoSs. Fitting both, the compactness and effective compactness relations to their results, they find that both approaches retain their approximate universality. They also find that the accuracy of the effective compactness relations is slightly better than the average density relations, as evidenced by their relatively smaller errors.

They also investigate the dependence of the product $\sigma\tau$ (i.e the product of f-mode frequency and the damping time) on the compactness. While this relation also shows strong universality, it does become independent of the compactness for $C \geq 0.2$, making this relation not that useful for inferring neutron star properties.

Analytical Treatment

Völkel and Kokkotas [VK19] analytically treat the problem of axial perturbations for spherical and non-rotating compact stars. With their novel approach, they provide an analytical explanation of the empirical relations discussed above: they find that the main fundamental mode is directly related to the potential on the surface of the star, and thus the relations that relate the fundamental mode frequency with the compactness of the star are well motivated. Only higher modes should depend and give insight into the actual EoS of the star.

2.1.2 Rotating Neutron Stars

With the emergence of more sophisticated numerical methods for the evolution of rotating compact star models, the evaluation of the universality for oscillation frequency and damping time relations has been made possible.

Note that, in rotating stars, perturbations can propagate in the prograde (co-rotating) or retrograde (counter-rotating) to the rotation of the neutron stars. For the f-mode, in particular, this causes a splitting of the observed f-mode frequency into co- and counter-rotating modes, respectively.

Cowling Approximation

Gaertig and Kokkotas [GK11] investigate, for the first time, approximately universal relations for the f-mode frequency and damping time for rapidly rotating stars. To this end, they use cold, polytropic EoS and perform perturbative calculations in the Cowling approximation (i.e. no spacetime perturbations).

For the f-mode frequency in the co-moving frame they find for the *stable* (cf. Section 3.4.1), co-rotating mode σ_c^s the relation

$$\frac{\sigma_c^s}{\sigma_0} = 1.0 - 0.27\left(\frac{\Omega}{\Omega_K}\right) - 0.34\left(\frac{\Omega}{\Omega_K}\right)^2 \tag{2.5}$$

where σ_0 is the f-mode frequency in the non-rotating case, Ω the angular rotation rate of the star, and Ω_K the Keplerian limit for the angular rotation rate. These relations appear highly accurate except close to the Keplerian limit. They also find equivalent relations for the counter-rotating f-mode.

To solve the inverse problem of determining stellar parameters from the mode frequency and damping time, universal expression for σ_0 and τ_0, the mode frequency and damping time in the non-rotating limit, are also formulated. The produced fits mirror prior results by Andersson and Kokkotas [AK98].

Doneva et al. [DGKK13] extend this analysis to higher order modes and for realistic EoS. They achieve similar approximately universal results, showing that the above approach can indeed be extended to higher order modes, and non-polytropic EoS. Their comparison to mode frequencies computed in non-linear, fully relativistic simulations without the Cowling approximation show that deviations mostly occur at angular rotation rates close to the Keplerian limit, justifying the use of the Cowling approximation for smaller angular rotation rates.

In a next step, Doneva and Kokkotas [DK15] consider different normalizations and independent variables in universal relations for rotating stars. Using the same numerical relativity simulations as above, they investigate the dependence of the f-mode frequencies and the corresponding damping times on the effective compactness η when normalized by the gravitational mass M_0 of the neutron star.

They find that, using the normalized rotational parameter $\hat{\Omega} = M_0\Omega$, the relation for the stable, co-rotating mode σ_c^{st} as measured by a co-moving observer, can be given by

$$M\sigma_c^{st} = \left(a_1 + a_2\hat{\Omega}\right) + \left(a_1 + a_2\hat{\Omega}\right)\eta \tag{2.6}$$

where, again, the a_i and b_i are coefficients derived from their best fit. Similar, η dependent relations are also formulated for the counter-rotating mode, and the respective damping times.

As such, the type of relations we previously had for non-rotating stars were also replicated for rotating neutron stars, albeit through rotation rate dependent relations.

Fully Relativistic, Without Approximation

Krüger and Kokkotas [KK20b, KK20a] finally depart from using approximations and investigate universal relations for the f-mode frequencies and their damping times using perturbation theory with dynamic spacetimes. They find qualitatively similar relations as in the Cowling approximation, with the changes in the coefficients being in line with expectation (due to the ignored spacetime perturbations in said approximation).

2.2 Tidal Deformability

In the following, we take a close look at the universal relations formulated in terms of the tidal deformability λ (cf. Section 3.3), i.e. the reaction of a compact star to an external tidal field. The tidal deformability has become particularly important in the study of neutron stars as it directly impacts the phase evolution of gravitational waves emitted e.g. by binary neutron star mergers [FH08]. As such, understanding which additional properties of neutron stars we can infer after extracting the tidal deformability from GW signals has become an important research endeavor.

Yagi and Yunes [YY17] initially introduce the I-Love-Q relation for cold, non-rotating neutron stars with polytropic equation of state: it relates the neutron star's normalized moment of inertia $\bar{I} = \frac{I}{M_0^3}$, its normalized quadrupole moment $\bar{Q} = \frac{Q}{M_0^3}$ and its normalized tidal deformability $\bar{\lambda} = \frac{\lambda}{M_0^5}$ (which is related to its *Love number* [Lov09] k_2 through $k_2 = \frac{3}{2}\frac{\lambda}{R^5}$, and hence the name of the relation). For slowly rotating stars with mass $M_\odot < M_0 < M_{max}$, where M_{max} is the maximum allowed mass by the considered EoS, they find an approximately universal relation of the form

$$\ln y_i = a x_o + b \ln x_i + c_i (\ln x_i)^2 + d_i (\ln x_i)^3 + e_i (\ln x_i)^4 \qquad (2.7)$$

for each $y_i, x_i \in \{\bar{I}, \bar{Q}, \bar{\lambda}\}$. This relation shows a low variance of $\sim 1\%$ across the considered EoSs.

While we currently lack a rigorous formalism to compute the tidal deformability of rapidly rotating neutron stars [PGMF18], investigations of the relation between $\bar{I}$ and $\bar{Q}$ have shown that approximate universality can be retained if a appropriately chosen rotation rate parameter (such as the dimensionless spin parameter χ) is assumed to be constant across all considered neutron stars [DYSK14, CDGS14, PA14].

For more complicated neutron star configurations, however, a partial breakdown of the universality of the I-(Love)-Q relation is observed: this includes analyses for magnetized neutron stars [HCPR14], dynamical tidal fields [MCF+13] and non-polytropic EoS [MMGF14, MOHN17], which show a significant increase of inter-EoS variance.

Similar relations have also been developed for higher orders of the tidal response of neutron stars, the so-called *multipole relations*. Since the effect of these higher moments on the gravitational wave phase evolution is greatly diminished, we will not further discuss these relations here. We instead refer the reader to the review by Yagi and Yunes [YY17] for a comprehensive discussion.

2.2.1 Relations Between Tidal Deformability and Oscillation Modes

Chan et al. [CSLL14] aim to to combine the $f-I$ relation by Tsui and Leung [TL05a] (as the effective compactness is given by $\eta = \sqrt{\frac{M^3}{I}}$) with the I-Love-Q relation by Yagi and Yunes [YY17] to investigate a potential f-Love relation between the f-mode frequency and the tidal deformability, not only for the quadrupolar $l = 2$ multipole, but also for higher multipole moments.

They find that for the l-th multipole, angular f-mode frequency ω_l and the l'-th multipole tidal deformability λ_l , there exists approximately universal relations only for $l = l'$, while for $l \neq l'$, the relations become EoS-dependent. For $l = l'$, the relation can be represented by

$$M\omega_l = a_0 + a_1 x + a_2 x^2 + a_3 x^3 + a_4 x^4 \tag{2.8}$$

where $x = ln\bar{\lambda}_l = ln\frac{\lambda_l}{M^{2l+1}}$ and the a_i are coefficients obtained through the best fit to their data.

The relative variance of this universal relation for incompressible stars is around 1%. As they also find that the f-mode frequencies of polytropic stars with polytropic index $n \leq 1$ deviate from the values for incompressible stars by at most 2%, they thus conclude that this relation has high universality across different polytropic EoSs.

In the same vein, Wen et al. [WLCZ19] also set out to find universal relations between the f-mode frequency and the tidal deformability of neutron stars. They find that, except for quark stars, the f-mode frequency and its damping time scale directly with the tidal deformability of the neutron stars, assuming a fixed mass, independently of the EoS. Consequently, if one of these can be accurately extracted from the GW signal (in addition to the mass of the neutron star), the other can constrained through their $f - \lambda$ relation. However, outside of a pictorial representation of their relation, Wen et al. fail to provide a numerical representation that includes the assumed mass of the neutron star, making it hard to quantify the accuracy of their results.

Most recently, Benitez et al. [BWG+20] again investigated the validity of the I-Love-Q relation for modern piecewise-polytropic EoSs. They are able to confirm its validity, and also find a close relation between I, λ and Q, and the of axial w-mode frequency of the neutron star that is universal across the EoSs considered in their study. However, as the w-mode frequencies of typical neutron stars lie outside of the frequency range of current and next-generation gravitational wave interferometers, the usefulness of this relation remains unclear.

2.3 Other Universal Relations

In addition to relations that involve the tidal deformability and oscillation modes of compact stars, work has also been put towards relations involving other stellar parameters. Such relations are important as the combination of these relations with the ones discussed above can allow us to estimate a wider range of neutron star parameters from one observation. The accuracy of such estimates, however, are of course limited by the accuracy of the original relations.

Lattimer and Prakash [LP01] investigate the stellar properties of neutron stars across different EoSs. In addition to finding relations between pressure, radius and mass for different classes of EoSs, they propose a universal relation for the total binding energy E_b which is defined by

$$E_b = M - \mathcal{N} m_n \tag{2.9}$$

where $\mathcal{N}$ is the total number of baryons and m_n the baryon mass. The relation is then given by

$$\frac{E_b}{M} \simeq 0.6 \frac{M}{R} \left(1 - 0.5 \frac{M}{R}\right)^{-1} \tag{2.10}$$

which, however, shows significant deviations for stiff EoSs.

Lattimer and Schutz [LS05] determine the moment of inertia I, mass and radius for neutron stars of different EoSs and find the approximately universal relation

$$I \simeq (0.237 \pm 0.008)\, M R^2 \left[1 + 4.2 \frac{M}{R} + 90 \left(\frac{M}{R}\right)^4\right] \tag{2.11}$$

which again shows significant deviations for stiff EoSs.

Steiner et al. [SLB16] and Breu and Rezolla [BR16] revisit these relations for a wider of modern, realistic EoSs and find similar relations with adjusted coefficients, for both the binding energy as well the moment of inertia.

Bozzola et al. [BSB18] investigate the stability of uniformly and differentially rotating compact stars using the turning point method by Friedman et al. [FIS88]. For a collection of both soft and stiff EoSs, they simulate neutron stars in various configurations and rotation profiles (uniform, 1- and 3-parameter differential rotation laws). They find three empirical, universal relations between the angular momentum J, gravitational mass M_0 and baryon mass M_b of the neutron star. These are given by

$$\frac{M_b}{M_b^*} = 1 + 0.51\left(\frac{J}{M_b^*}\right)^2 - 0.28\left(\frac{J}{M_b^*}\right)^4 \tag{2.12}$$

$$\frac{M_b}{M_b^*} = 0.93\frac{M_0}{M_0^*} + 0.07 \tag{2.13}$$

$$\frac{M_0}{M_0^*} = 1 + 0.29\left(\frac{J}{M_0^*}\right)^2 - 0.10\left(\frac{J}{M_0^*}\right)^4 \tag{2.14}$$

where M_b^* and M^* are the maximum baryon and gravitational mass of a TOV star. These relation appear highly accurate with relative error in the order of 1%. However, strange star EoSs show a significant deviation from these relations.

Gao et al. [GAC+20] relate the baryon mass M_b and gravitational mass M_0 of neutron stars across different EoSs and find approximate universality: they evolve models of (non)-rotating neutron stars using `rns` [SF95, NSGE98] and find for non-rotating neutron stars the quadratic relation

$$M_b = M_0 + \frac{M_0^2}{R_{1.4}} \tag{2.15}$$

where $R_{1.4}$ is the radius of a neutron star with mass $1.4M_\odot$. This relation shows a maximum relative error of 1.8%. Coughlin et al. [CDK+17] previously put forward the relation

$$M_b = M_0 + aC^n M_0 \tag{2.16}$$

with fitting parameters a and n. This relation has similar maximum relative error, but depends on the compactness C. In comparison, Gao et al.'s proposal allows for a direct determination of $R_{1.4}$ if M_b and M_g are determined independently.

For uniformly rotating neutron stars with normalized spin period $\bar{P} = \frac{P}{P_K}$ where P_K is the spin period at the Keplerian limit, Gao et al. find the more involved relation

$$M_b = M_0 + \frac{M_0^2}{R_{1.4}e^{\frac{1}{4\bar{P}}}} \tag{2.17}$$

which achieves a maximum relative error of 3.3%.

2.4 Binary Neutron Stars

We now consider the case of BNS mergers: Bauswein and Janka [BJ12, BJHS12] perform a first study into the relation between the post-merger peak frequency f_{peak} (which is related for the f-mode frequency σ, as discussed above) and the stellar properties of the post-merger differentially rotating object of a symmetric (i.e. equal mass) binary neutron star merger. To this end, they simulate the binary coalescence for a wide set of tabulated, microphysical EoSs, and extract f_{peak} in cases where the merger results in a long-lived neutron star remnant. Comparing f_{peak} to characteristic quantities of the different EoSs, they find a number of empirical relations, in particular the relation

$$f_{peak} = \begin{cases} -0.2823R_{1.6} + 6.284 & \text{for } f_{peak} < 2.8\text{kHz} \\ -0.4667R_{1.6} + 8.713 & \text{for } f_{peak} \geq 2.8\text{kHz} \end{cases} \tag{2.18}$$

between the post-merger peak frequency f_{peak} and the radius $R_{1.6}$ of a neutron star with mass $M_0 = 1.6M_{\odot}$.

2.4.1 Binary Love Relations

Yagi and Yunes [YY16] investigate universal relations for BNSs that involve the tidal deformabilities of the pre-merger neutron stars. Given each star's respective dimensionless tidal deformabilities $\bar{\lambda}_1 = \frac{\lambda_1}{M_1^5}$ and $\bar{\lambda}_2 = \frac{\lambda_2}{M_2^5}$, they attempt to characterize the tidal properties of the whole binary through their symmetric and anti-symmetric combinations

$$\bar{\lambda}_s = \frac{\bar{\lambda}_1 + \bar{\lambda}_2}{2} \qquad \bar{\lambda}_a = \frac{\bar{\lambda}_1 - \bar{\lambda}_2}{2} \tag{2.19}$$

Through simulation of BNSs with different realistic EoSs, they observe that these two quantities have an approximately universal relation with deviations up to 10% for small mass ratios $q \leq 0.5$ and large values of $\bar{\lambda}_s$. However, for even mass ratios $q \approx 1$ and small $\bar{\lambda}_s$, significantly larger deviations are observed.

2.4.2 Effective Tidal Deformability and Chirp Mass

The direct observation of a BNS merger allows us to accurately measure its *effective tidal deformability* [FH08]

$$\tilde{\lambda} = (11M_2 + M)\,\bar{\lambda}_1 M_1^4 + (11M_1 + M)\,\bar{\lambda}_2 M_2^4 \tag{2.20}$$

from the phase evolution of the GW signal. Raithel et al. [RÖP18] find that, if we are given the *chirp mass* $\mathcal{M}$, defined by

$$\mathcal{M} = \frac{(M_1 M_2)^{\frac{3}{5}}}{(M_1 + M_2)^{\frac{1}{5}}} = M_1 \frac{q^{\frac{3}{5}}}{(1+q)^{\frac{1}{5}}} \tag{2.21}$$

the effective tidal deformability is approximately independent of the individual neutron stars' masses and can instead be used to directly measure their radii. Their analytical derivations in the Newtonian limit with metric corrections yield the relation

$$\tilde{\lambda} = \frac{15 - \pi^2}{3\pi^2} \xi^{-5} (1 - 2\xi)^{\frac{2}{5}} \tag{2.22}$$

with

$$\xi = \frac{2^{\frac{1}{5}} \mathcal{M}}{R} \tag{2.23}$$

for $q \approx 1$, independently of EoS. The individual masses of the neutron stars only enter at order $O((1-q)^2)$, which leads to an error of $\sim 4\%$ when estimating the stars' radii while neglecting their masses.

Zhao and Lattimer [ZL18] relate the tidal deformability ratio $\frac{\bar{\lambda}_1}{\bar{\lambda}_2}$ to the mass ratio q of a BNS, showing that, for a chirp mass $\mathcal{M} \leq 1.4 M_\odot$ with it holds that

$$\frac{\bar{\lambda}_1}{\bar{\lambda}_2} \sim q^6 \tag{2.24}$$

for the full range of parametric, piecewise polytropic EoSs through numerical relativity simulations of all possible parameter configurations. They also find that, in this case, the tidal deformability ratio can be bounded by

$$q^{n_-} \leq \frac{\bar{\lambda}_1}{\bar{\lambda}_2} q^{n_{0+} + q n_{1+}} \tag{2.25}$$

where the exponents n_-,n_{0+} and n_{1+} depend on the system's chirp mass $\mathcal{M}$. By establishing such bounds, the a priori uncertainty on the tidal deformability used for the evaluation of gravitational wave signals can be tremendously reduced (if one knows the mass ratio of the BNS), allowing for a more accurate parameter estimation.

2.4.3 Tidal Coupling Constant

Bernuzzi et al. [BDN15] propose the *tidal coupling constant* κ_2^t as tidal parameter in universal relations for BNS mergers. It is given by

$$\begin{aligned}
\kappa_2^t &= 2\left(\frac{q^4}{(1+q)^5}\frac{k_2^1}{C_1^5} + \frac{q^4}{(1+q)^5}\frac{k_2^2}{C_2^5}\right) \\
&= 2\left[q\left(\frac{M_1/M}{M_1/R_1}\right)^5 k_2^1 + \frac{1}{q}\left(\frac{M_2/M}{M_2/R_2}\right)^5 k_2^2\right] \\
&= 2\left[q\left(\frac{R_1}{M}\right)^5 k_2^1 + \frac{1}{q}\left(\frac{R_2}{M}\right)^5 k_2^2\right]
\end{aligned} \tag{2.26}$$

where C_i, M_i, R_i and k_2^i are the compactnesses, masses, radii and Love numbers of the neutron stars in the BNS. It combines the tidal deformabilities of the individual pre-merger stars to produce a combined tidal parameter, and emerges during the modeling of the binary neutron star merger using the effective-one-boy approach [BNB+14]. Similarly to the effective tidal deformability, it affects the phase evolution the gravitational wave signal emitted during a BNS merger.

To develop their relations using κ_2, they consider (3+1)D numerical relativity simulations of irrotational neutron star binaries of unequal mass up to 30ms after the merger, and only consider mass configuration that result in a long-lived neutron star remnant.

They find for each, the reduced binding energy E_b^{mrg}, the reduced angular momentum j_{mrg} and the normalized gravitational wave frequency $M_0\omega_{mrg}$ at the time of the merger, and the primary, normalized peak $M_0 f_{peak}$ of the post-merger signal, an approximately universal relation of the form

$$X(\kappa_2^t) = X_0\frac{1 + n_1\kappa_2^t + n_2\left(\kappa_2^t\right)^2}{1 + d_1\kappa_2^t} \tag{2.27}$$

where the coefficients X_0, n_1, n_2 and d_1 are found through the best fit to their numerical data.

Bernuzzi et al. consequently argue that κ_2^t should be considered as the principal tidal parameter that will allow for the modeling and analysis of pre- and post-merger gravitational wave signals detected from BNS mergers. Inspired by these results, a number of other works has also been put forward that investigate similar relations for a wider range of BNS configurations [ZBR+18, RT16].

2.4.4 High Resolution Simulations

Kiuchi et al. [KKK+20] simulate the merger of 26 non-spinning BNS (i.e. the pre-merger stars have no individual spins) with improved grid resolution using their SACRA-MPI numerical relativity code. They extract the gravitational wave signal from these simulations and investigate whether previously proposed universal relations [BJ12, RBC+13, RT16, ZBR+18] still hold in this high resolution, fully relativistic setting.

Their findings reveal systematic errors in the relations produced by earlier work: as already explained in [RT16], mass ratios $q \leq 0.8$ might cause significant deviations from the universal relations presented therein. While such mass ratios were deemed unrealistic in [RT16], the observation of GW170817 [AoLV17] has shown that $q \leq 0.8$ is indeed a possibility.

Kiuchi et al. therefore investigate potentially new relations across a wider range of mass ratios: they propose new universal relations for the pre-merger peak frequency f_{peak}, its amplitude h_{peak}, and the post-merger f_2 peak that depend on the *reduced* (or binary) *tidal deformability* [Fav14]

$$\tilde{\Lambda} = 32\frac{\tilde{\lambda}}{M^5} = \frac{8}{13}\left[\left(1 + 7\zeta - 31\zeta^2\right)\left(\bar{\lambda}_1 + \bar{\lambda}_2\right) - \sqrt{1-4\zeta}\left(1 + 9\zeta - 11\zeta^2\right)\left(\bar{\lambda}_1 - \bar{\lambda}_2\right)\right] \tag{2.28}$$

where $\zeta = \frac{M_1 M_2}{(M_1+M_2)^2} = \frac{q}{(1+q)^2}$ is the *symmetric mass ratio* of the BNS. Their relations take the form

$$X = c_1 + c_2\tilde{\Lambda}^{\frac{1}{5}} \tag{2.29}$$

where X is a dimensionless, mass-normalized variant of one of the above mentioned quantities, and the coefficients c_1 and c_2 depend on the symmetric mass ratio ζ.

While this new relation marks an improvement to previously proposed ones, it still incurs a relative error of $3 - 10\%$, and Kiuchi et al. point out that more work has to be put towards including all physically relevant processes that might occur during the merger into future simulations.

Based on the same numerical data, Yamasaki et al. [YTK20a] recently put forward a universal relations of the same form as above, but involving the normalized spin period $\frac{P}{M}$ of a potential, long-lived post-merger remnant neutron star: as this spin period might be observed through (electromagnetic) channels independent from any GW observations of the BNS merger, this might allow us to put additional constraints on our priors for the binary tidal deformability, thus allowing us to improve our estimates thereof.

2.5 Discussion

Current and next generation gravitational wave detectors will be able to detect GWs emitted from BNS mergers and potentially from the post-merger phase. Gravitational waves emitted from single neutron stars, however, will most likely remain outside of our observational capabilities for the immediate future. As such, extending the work on universal relations for BNSs presents an important direction of work that will enable us to further expand the amount of information we can obtain from observing GW signals of BNS mergers.

As pointed out by Kiuchi et al. [KKK+20], however, a comprehensive treatment of all relevant physical mechanisms that are involved in a BNS merger is still required to obtain accurate, approximately universal relations. However, the treatment of BNS mergers through numerical relativity simulations remains very costly.

We thus propose, in this thesis, a perturbative approach to developing such universal relations, assuming that the merger results in a long-lived remnant neutron star: guided by general constraints on, e.g., the total mass and angular moment of the pre- and post-merger phases of a BNS merger obtained through numerical relativity simulations, we can individually treat the neutron stars in these phases and compare their properties to obtain new relations. In particular, we can leverage the work on universal relations for single, rapidly rotating neutron stars (e.g. [KK20b]) to extend universal relations obtained through numerical relativity simulations.

Theory

3

In the following, we outline the theory underlying the work presented in this thesis. This mainly involves the formulation of neutron star equilibrium models for non-rotating neutron stars and numerically solving the corresponding TOV equations, as well as the computation of their tidal deformabilities. We will also briefly introduce the concept of quasi-normal modes of neutron stars which will be important for our subsequent analyses.

3.1 Neutron Star Equilibrium Model and TOV Equations

We consider the following metric for a non-rotating neutron star

$$ds^2 = -e^{\nu(r)}dt^2 + e^{\lambda(r)}dr^2 + r^2\left(d\theta^2 + sin^2\theta d\phi^2\right) \tag{3.1}$$

where the function $\lambda(r)$ fulfills the conditions

$$e^{\lambda(r)} = \left(1 - \frac{2m(r)}{r}\right)^{-1} \quad \text{with} \quad m(r) = \int_0^r 4\pi\bar{r}^2\epsilon(\bar{r})d\bar{r} \tag{3.2}$$

With the energy-momentum tensor of a perfect fluid

$$T_{\mu\nu} = (\epsilon + p)u_\mu u_\nu + pg_{\mu\nu} \tag{3.3}$$

the metric can be determined by solving the Tolman-Oppenheimer-Volkoff (TOV) equations

P. Manoharan, *Universal Relations for Binary Neutron Star Mergers with Long-lived Remnants*, BestMasters, https://doi.org/10.1007/978-3-658-36841-8_3

$$\frac{dm}{dr} = 4\pi r^2 \epsilon(r) \tag{3.4}$$

$$\frac{dp}{dr} = -(\epsilon + p)\frac{m + 4\pi r^3 p}{r(r-2m)} \tag{3.5}$$

$$\frac{d\nu}{dr} = 2\frac{m + 4\pi r^3 p}{r(r-2m)} \tag{3.6}$$

through integration from the center of the neutron star to its surface. We determine the surface to be at the radius R at which the pressure first reaches $p(R) = 0$.

For the initial values, we set $m(0) = 0$ (and thus $e^{\lambda(0)} = 1$). Given a *central energy density* $\epsilon_c = \epsilon(0)$, the corresponding central pressure $p(0)$ can be determined through the *equation of state* that we assume for the neutron star, which we will discuss below.

We do not require an initial value for $\nu(r)$ as its exact value is irrelevant for our use case: we are only interested in the final mass $M_0 = m(R)$, radius R and the mass and pressure functions $m(r)$ and $p(r)$. We will only require an initial value for $\frac{d\nu}{dr}$ when we compute the Love number of the neutron star (cf. Section 3.3), but this is already given by the other initial values.

Since we aim to evolve neutron stars for a given mass $M_0 = m(R)$, we need to specify the correct central energy density ϵ_c that results in this mass. For low to medium mass neutron stars, we use a simple bisection search to find the correct central density. Closer to the maximum mass limit, the bisection search usually fails, and we fall back to linearly searching through the full range of central densities.

Equation of State

For the equation of state (EoS) that we need to solve the TOV equations above, we utilize piecewise-polytropic approximations, proposed by Read et al. [RLOF09], of the five phenomenological EoSs `SLy` [DH01], `WFF1` [WFF88], `APR4` [APR98], `H4` [LNO06] and `MS1` [MS96].

These piecewise-polytropic approximations are defined as follows: for a given set of rest-mass densities thresholds $\rho_0 < \rho_1 < \rho_2$, the pressure density relations of these EoSs are given by

$$p(\rho) = K_i \rho^{\Gamma_i} \quad \rho_{i-1} \leq \rho < \rho_i \tag{3.7}$$

where Γ_i is the adiabatic index, and K_i the proportionality constant specific to the given EoS.

Now, by the first law of thermodynamics, the energy density ϵ and rest mass density ρ are related by

$$d\frac{\epsilon}{\rho} = -p d\frac{1}{\rho} \tag{3.8}$$

Integrating this equation and solving for ϵ, we obtain for each piecewise-polytropic EoS the expression

$$\epsilon(\rho) = (1 + a_i)\rho + \frac{K_i}{\Gamma_i - 1}\rho^{\Gamma_i} \tag{3.9}$$

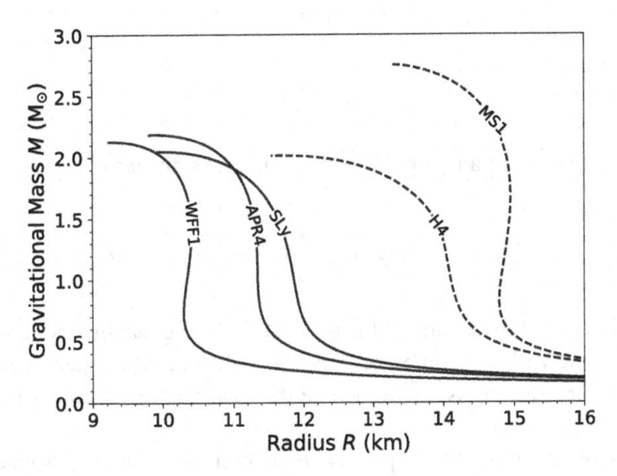

Figure 3.1 The mass-radius relation for non-rotating stars of the five phenomenological EoSs considered in this thesis. EoSs that we consider to be *soft* are indicated by a solid line, and those we consider *stiff* by a dotted dashed line

where the a_i are EoS specific constants. The values for the constants K_i, Γ_i and a_i for the five EoS that we use in this thesis are given in tabulated form in [RLOF09, Table III]. Given these two relations, we can thus determine the central pressure $p(0)$ given a central energy density, and evolve the TOV equations described above.

We integrate the TOV equations using the ODE solver `solve_ivp` implemented in the `Python` package `Scipy`, which internally implements the explicit 5th-order Runge-Kutta method. The resulting mass-radius relations for the five EoSs considered in this thesis are given in Figure 3.1. To help discussions in the following sections, we partition these EoSs into two classes: *soft* (WFF1, APR4 and SLy) and *stiff* (H4 and MS1).

3.2 Neutron Star Bulk Properties

Throughout this thesis, we will consider several bulk properties of neutron stars. We briefly define them here. As mentioned above, we define the *radius* R of a neutron star to be the distance from the center of the neutron star at which the pressure reaches zero, i.e. $p(R) = 0$. Subsequently, the *mass*, or *gravitational mass*, M_0 of a non-rotating neutron star is given by

$$M_0 = m(R) = \int_0^R 4\pi \bar{r}^2 \epsilon(\bar{r}) d\bar{r}. \tag{3.10}$$

and its *compactness* C by

$$C = \frac{M_0}{R}. \tag{3.11}$$

The *baryon mass* M_b of a non-rotating neutron star is given by

$$M_b = \int_0^R 4\pi r^2 \rho(r) \left(1 - \frac{2m(r)}{r}\right)^{-\frac{1}{2}} dr \tag{3.12}$$

where ρ is the rest mass density of the neutron star. Note that in other publications, the baryon mass will sometimes be denoted with M_0, which here stands for its gravitational mass as we use M to denote the total mass of a binary neutron star system.

The *moment of inertia* I of a spherical body uniformly rotating around, e.g., the z axis, in flat space is given by

$$I = \int_V \left(r^2 - z^2\right) \rho(\vec{r}) dV \tag{3.13}$$

However, this expression cannot be straightforwardly generalized for rotating neutron stars (such as the remnant of a binary neutron star merger) [PS17], and in this case the moment of inertia is typically defined by

$$I = \frac{J}{\Omega} \tag{3.14}$$

where Ω is the angular rotation rate of the neutron star, and J its *angular momentum*. Since we directly use the data from [KK20b] for the moment of inertia, and do not compute it ourselves, we refer to [PS17] for a comprehensive treatment of the equilibrium models of rotating neutron stars, and their properties, including the angular momentum and moment of inertia.

Given the moment of inertia I and mass M_0 of a neutron star, its *effective compactness* η is a dimensionless quantity that is given by

$$\eta = \sqrt{\frac{M_0^3}{I}} \tag{3.15}$$

As we saw in Chapter 2, the effective compactness has been shown to be a useful alternative to other bulk properties such as the compactness or moment of inertia in the formulation of universal relations.

3.3 Tidal Deformability of Neutron Stars

The tidal Love number k_2 is named after Augustus Edward Hough Love [Lov09], who investigated the deformation of the earth under the influence of the moon's gravitational attraction. In a similar fashion, we can investigate the reaction of a neutron star to an external gravitational field sourced by another body, e.g. another neutron star (as is the case for BNSs). We here give a basic description of the *quadrupolar tidal deformability* following the discussion by Hinderer [Hin08, Hin09]. A more general discussion for higher order responses of a compact star to tidal fields is given by Damour and Nagar [DN09], and Binnington and Poisson [BP09]. A nice and detailed discussion of this topic with extensions to surface deformations can also be found in the master thesis by Landry [Lan14].

3.3.1 The Tidal Love Number

In the Newtonian limit, the quadrupolar tidal field $\mathcal{E}_{ij}$ induced by an external gravitational potential Φ_{ext} is given by

$$\mathcal{E}_{ij} = \frac{\partial^2 \Phi_{ext}}{\partial x^i \partial x^j} \tag{3.16}$$

In a primary body, e.g. a neutron star, this tidal field induces density perturbations $\delta\rho$ due to the interactions of the body's own gravitational potential with the external tidal field. These density perturbations result in an *induced quadrupole moment* Q_{ij} given by

$$Q_{ij} = \int d^3x\, \delta\rho(\vec{x}) \left(x_i x_j - \frac{1}{3} r^3 \delta_{ij} \right) \quad (3.17)$$

At first order, this response depends linearly on the tidal field $\mathcal{E}_{ij}$ through

$$Q_{ij} = -\frac{2}{3} k_2 R^5 \mathcal{E}_{ij} \quad (3.18)$$

where k_2 is the *tidal Love number*. The index 2 here indicates the we here only consider the response of the neutron star to the external tidal field due to $l = 2$ perturbations of the neutron star (cf. Section 3.4), which have the strongest coupling to gravitational wave emission [Sch08], and are thus of most interest to us.

We can simplify Equation (3.18) by introducing the *quadrupolar tidal deformability* λ defined by

$$\lambda = \frac{2}{3} R^5 k_2 \quad (3.19)$$

leading to

$$Q_{ij} = -\lambda \mathcal{E}_{ij} \quad (3.20)$$

Assuming that the tidal field is axisymmetric along the x_3 axis, the elements of the tidal field tensor and quadrupole moment tensor are given by [Tho98]

$$\mathcal{E}_{11} = \mathcal{E}_{22} = -\frac{1}{2} \mathcal{E}_{33} = \mathcal{E} \qquad Q_{11} = Q_{22} = -\frac{1}{2} Q_{33} = -Q \quad (3.21)$$

and the relation above further reduces to

$$Q = -\lambda \mathcal{E} \quad (3.22)$$

In general relativity, the expressions in Equations (3.16) and (3.17) are no longer valid. Instead, more involved expressions for the mass multipole moments in curved space (as given by e.g. Geroch [Ger70]) are required, and the tidal field needs to be defined through the Riemann tensor of the external body, leading to, in particular, two different types of tidal Love numbers [DN09].

For the case of the quadrupolar tidal response of the central body, however, these calculations still again yield the same relation as in Equation (3.20). We will therefore focus our attention here on how we can compute the tidal Love number, and refer the reader to [DN09, BP09, Lan14] for a more detailed discussion of the different types of tidal responses in general relativity.

3.3.2 Computing the Tidal Love Number

Consider a spherically symmetric neutron star with the metric $g^0_{\mu\nu}$ described in Equation (3.1). When this star experiences static perturbations induced by a static, external tidal field, its new metric can be written, at linear order, as

$$g_{\mu\nu} = g^0_{\mu\nu} + h_{\mu\nu} \tag{3.23}$$

where the $h_{\mu\nu}$ are the perturbations of the metric. For static $l = 2$ perturbations (cf. Section 3.4), and using the Regge-Wheeler gauge [RW57], all non-diagonal elements of the perturbation vanish and it is given by [TC67]

$$h_{\mu\nu} = \text{diag}\left[e^{-\nu(r)}H_0(r), e^{\lambda(r)}H_2(r), r^2K(r), r^2\sin^2\theta K(r)\right]Y_{2m}(\theta,\phi) \tag{3.24}$$

where Y_{lm} are spherical harmonics. Combining this with the perturbed energy-momentum tensor $\delta T_{\mu\nu}$ of the star, and the linear, perturbed Einstein equations

$$\delta G_{\mu\nu} = 8\pi\delta T_{\mu\nu} \tag{3.25}$$

we find that $H_2 = H_0 \equiv H(r)$, and the differential equation for H

$$\begin{aligned} &H'' + H'\left[\frac{2}{r} + e^{\lambda(r)}\left(\frac{2M(r)}{r^2} + 4\pi r(p-\epsilon)\right)\right] \\ &+ H\left[-\frac{6e^{\lambda(r)}}{r^2} + 4\pi e^{\lambda(r)}\left(5\epsilon + 9p + \frac{\epsilon+p}{dp/d\epsilon}\right) - \left(\frac{d\nu}{dr}\right)^2\right] = 0 \end{aligned} \tag{3.26}$$

The general solution for H can be given as

$$H(r) = c_1 Q_2^2\left(\frac{r}{M_0} - 1\right) + c_2 P_2^2\left(\frac{r}{M_0} - 1\right) \tag{3.27}$$

where Q_2^2 and P_2^2 are the *associated Legendre Polynomials*. Inserting the expressions for these polynomials, and analyzing the behavior of H at large r, we get that

$$H(r) = \frac{8}{5}\left(\frac{M_0}{r}\right)^3 c_1 + O\left(\left(\frac{M_0}{r}\right)^4\right) + 3\left(\frac{r}{M_0}\right)^2 c_2 + O\left(\frac{r}{M_0}\right) \tag{3.28}$$

where c_1 and c_2 are two constant coefficients.

Now, for a spherically symmetric star with quadrupole moment Q_{ij} in an external tidal field $\mathcal{E}_{ij}$, the metric coefficient g_{tt} at large r is given by [TH85]

$$\frac{(1-g_{tt})}{2} = -\frac{M_0}{r} - \frac{3}{2}\frac{Q_{ij}}{r^5}\left(r^i r^j - \frac{1}{3}\delta^{ij}\right) + O\left(\frac{1}{3}\right) + \frac{1}{2}\mathcal{E}_{ij}x^i x^j + O(r^3) \tag{3.29}$$

for the region outside the neutron star in which the external tidal field can be considered uniform.

To obtain the value of g_{tt}, we insert the solution of H from Equation (3.28) into Equation (3.24). We can then insert the g_{tt} component into Equation (3.29), and obtain for the coefficients c_1 and c_2 (using Equation (3.22))

$$c_1 = \frac{15}{8}\frac{1}{M_0^3}\lambda\mathcal{E} \qquad c_2 = \frac{1}{3}M_0^2\mathcal{E} \tag{3.30}$$

Inverting the expression for H in Equation (3.28) with these coefficients to solve for λ, we finally get that the tidal Love number k_2 of an irrotational neutron star with compactness C is given by

$$\begin{aligned} k_2 =& \frac{8C^5}{5}\left(1-2C^2\right)\left[2+2C(y-1)-y\right] \\ &\times \{2C\left[6-3y+3C\left(5y-8\right)\right] \\ &\quad +4C^3\left[13-11y+C(3y-2)+2C^2(1+y)\right] \\ &\quad +3(1-2C^2)\left[2-y+2C(y-1)\right]\log(1-2C)\}^{-1} \end{aligned} \tag{3.31}$$

where $y = R\frac{H'(R)}{H(R)}$.

Our main computational task in this thesis will be to integrate Equation (3.26) from $r = 0$ to R to obtain the values for $H(R)$ and $H'(R)$. As initial values for H and H' we follow the suggestion by Damour and Nagar [DN09] of $H(r_0) = r_0^2$ and $H'(r_0) = 2r_0$ for some small initial radius $r_0 \sim 10^{-6}$.

In order to integrate Equation (3.26), we first transform the second order ODE into a coupled system of first order ODEs

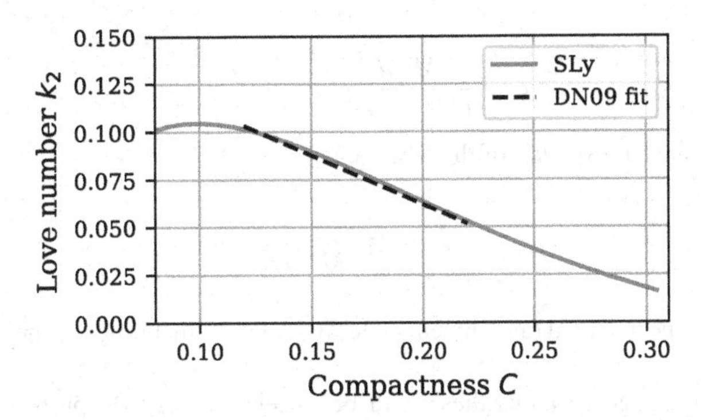

Figure 3.2 The relation between the compactness C and Love number k_2 for the EoS SLy, together with the linear fit proposed by Damour and Nagar [DN09]

$$
\begin{aligned}
G &= H' \\
G' &= -G\left[\frac{2}{r} + e^{\lambda}\left(\frac{2m}{r^2} + 4\pi r(p-\epsilon)\right)\right] \\
&\quad - H\left[-\frac{6e^{\lambda}}{r^2} + 4\pi e^{\lambda}\left(5\epsilon + 9p + \frac{\epsilon + p}{dp/d\epsilon}\right) - \left(\frac{d\nu}{dr}\right)^2\right]
\end{aligned}
\tag{3.32}
$$

which we solve simultaneously with the TOV equations described in Section 3.1.

Exemplarily, we illustrate the resulting Love numbers for the EoS SLy depending on the compactness C, together with the linear fit proposed by Damour and Nagar [DN09, Equation (116)] (for $0.12 \leq C \leq 0.22$) in Figure 3.2.

3.3.3 Binary Tidal Deformability

Our main focus in this thesis will be on the treatment of binary neutron star mergers. In this context, the principal tidal deformability parameter is the *binary tidal deformability* $\tilde{\Lambda}$ [Fav14] of a BNS before the merger: for a BNS of mass ratio q and total mass $M = M_1 + M_2$, such that $q = \frac{M_1}{M_2} \leq 1$, $\tilde{\Lambda}$ is given by [Fav14]

$$
\begin{aligned}
\tilde{\Lambda} =& \frac{8}{13}\left[\left(1 + 7\zeta - 31\zeta^2\right)\left(\bar{\lambda}_1 + \bar{\lambda}_2\right)\right. \\
&\left. - \sqrt{1 - 4\zeta}\left(1 + 9\zeta - 11\zeta^2\right)\left(\bar{\lambda}_1 - \bar{\lambda}_2\right)\right]
\end{aligned}
\tag{3.33}
$$

where

$$\zeta = \frac{M_1 M_2}{(M_1 + M_2)^2} = \frac{q}{(1+q)^2} \tag{3.34}$$

is the *symmetric mass ratio* of the BNS, and

$$\bar{\lambda}_i = \frac{\lambda_i}{M_i^5} \tag{3.35}$$

are the mass-normalized (and dimensionless) tidal deformabilities of the individual stars.

The primary goal of this thesis will be to relate $\tilde{\Lambda}$ to bulk properties of the post-merger remnant.

3.4 Quasi-normal Modes of Neutron Stars

We next give a brief description of the quasi-normal modes of neutron stars. For a comprehensive discussion, we refer the reader to [KS99].

The oscillation of a classical, bound oscillator can be described as a superposition of its normal modes, e.g. the displacement y of a linear oscillator is given by

$$y(t,x) = \sum_{n=1}^{\infty} a_n e^{i\omega_n t} y_n(x) \tag{3.36}$$

Similarly, the damped oscillations of a perturbed neutron star (as its oscillations will gradually lose energy over time, e.g. in the form of gravitational waves) can be described by the superposition of its *quasi-normal modes* [KS99] $\chi_{lm}(t,r)$ through

$$\chi(t,r,\theta,\phi) = \sum_{lm} \frac{\chi_{lm}(t,r)}{r} Y_{lm}(\theta,\phi) \tag{3.37}$$

where Y_{lm} are the tensor spherical harmonics.

The value of the quasi-normal mode $\chi_{lm}(t,r)$ depends on the perturbation $h_{\mu\nu}$ of the metric of the neutron star. As such, depending on the type of the initial perturbation of the star, we will have different *complex* frequencies $\omega_{lm} = \omega_{lm}^r + i\omega_{lm}^i$ associated with the quasi-normal modes of the resulting oscillations. The real part ω_{lm}^r of these frequencies describe the actual frequency of the oscillation, often simply denoted by ω, whereas the imaginary part describes the damping, or decay, of the oscillation (e.g. through dissipation), and is sometimes also denoted as the *damping time* τ of this quasi-normal mode.

A large number of works, initiated by the seminal paper by Throne and Campolattaro [TC67], has focused on analyzing and understanding the rich and varied oscillation spectrum of perturbed neutron stars. While many different families of oscillation modes, associated with different types of perturbations, exist, we will throughout this thesis focus on the $l = 2$ modes due to their strong coupling to gravitational wave emission [Sch08]. In particular, we will focus on the $l = 2$ f-mode, or fundamental pressure mode, that is associated with non-radial polar oscillations with pressure as the restoring force. This mode is important for us due to its strong coupling to external tidal fields [HL99], i.e. it will get easily excited during astrophysical processes like the merger of two binary neutron stars. For a typical neutron star, the f-mode has a frequency in the range of $1.5 - 3$kHz.

For non-rotating stars, the f-mode frequency is independent of the index m due to the spherical symmetric of the underlying spacetime. As such, we will denote the f-mode frequency simply by

$$\sigma = \frac{\omega}{2\pi} \tag{3.38}$$

where ω is the real, angular frequency of the $l = 2$ f-mode. Note that both, the frequency σ and the angular frequency ω, will be used throughout this thesis whenever appropriate.

3.4.1 Quasi-normal modes of Rotating Neutron Stars

For rotating neutron stars, the degeneracy of the index m for the quasi-normal modes vanishes: for the f-mode, we observe a splitting of the oscillations frequencies, essentially caused by the f-mode oscillations traveling in direction of the neutron star's rotation ($m < 0$) or against the direction of rotation ($m > 0$) [PS17].

Depending on whether we consider an inertial or co-moving observer, the split oscillation modes will develop differently: for an inertial observer, the co-rotating mode will show a higher frequency with increasing rotation rate of the star than the counter-rotating mode, whereas exactly the opposite is true for the co-moving observer (i.e. counter-rotating modes have higher frequency).

The split f-mode frequencies σ_i for $l = |m| = 2$ as observed from an inertial observer is illustrated in Figure 3.3 (taken from [KK20b]). It shows the normalized frequency $\frac{\sigma_i}{\sigma_0}$ against the normalized angular rotation rate $\frac{\Omega}{\sigma_0}$ of the star, where σ_0 is the f-mode frequency of the corresponding non-rotating star.

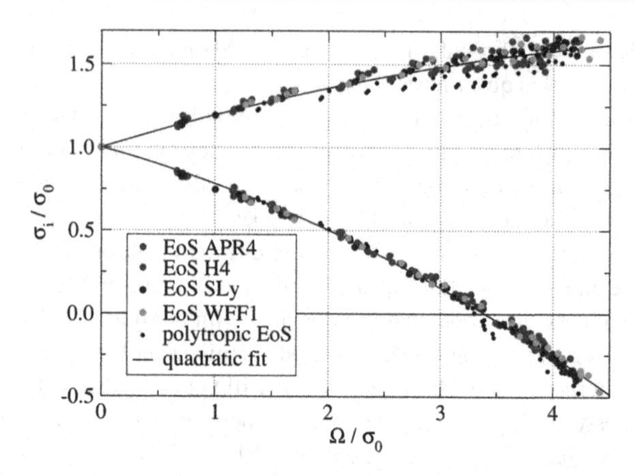

Figure 3.3 The splitting of the f-mode frequency in rotating stars, as observed from an inertial observer [KK20b, Figure 1]

As we can see, for the inertial observer, the counter-rotating mode can pass the zero frequency threshold, reaching negative frequencies and leading to the well-known Chandrasekhar-Friedman-Schutz (CFS) instability [Cha70, FS78]. As such, the counter-rotating mode is also referred to as the *potentially unstable* mode, whereas the co-rotating mode is referred to as the *stable* mode.

Throughout this thesis, we will refer to the frequency of the stable, co-rotating mode with σ^s, whereas the potentially unstable, counter-rotating mode will be denoted by σ^u (we only consider the $l = |m| = 2$ modes).

3.5 Goodness of Fit and Error Estimates

The general goal of all of our analyses will be to produce functional approximations of the relation between two parameters of a single neutron stars, or a BNS merger. To produce these relations, we first produce scatter plots of the two considered quantities, often with a third parameter (rotation rate of the star) encoded in the color of each point.

After considering these scatter plots, we will try to guess a function F that relates the two quantities X and Y, i.e.

$$Y = F(X) \tag{3.39}$$

with high accuracy. This function will typically be a linear function, although we also consider higher order polynomials (for the sake of brevity, we will only present the best fits throughout this thesis). The coefficients of the best fit for F are found by minimizing the squared error. In our case, we use the optimization function `curve_fit` and `polyfit` implemented in the Python packages `scipy` and `numpy`.

To estimate the goodness of our fit, we will compare the estimates $(X_i, F(X_i))$ produced by F to the actual data points (X_i, Y_i) and compute the root-mean-squared error (or RMSE) $\overline{E}$

$$\overline{E} = \sqrt{\frac{1}{n}\sum_i (F(X_i) - Y_i)^2} \tag{3.40}$$

where n is the number of data points we consider. Ideally, we want to minimize this error. To get a better feeling for the error, we will also consider the average relative error $\bar{e}$

$$\bar{e} = \frac{1}{n}\sum_i \frac{|F(X_i) - Y_i|}{|Y_i|} \tag{3.41}$$

While there exist more sophisticated measures to assess goodness of fit and accuracy (e.g. the χ^2 or R^2 statistics), these require a priori assumptions about the underlying distribution of the data points, which we cannot make easily: on the one hand, this is due to the multivariate nature of our data. On the other hand, this is also a result of our data coming from numerical computations, and not repeated empirical observations. If applicable to our use case, these statistics mostly reduce to the RMSE as well. As such, we remain with the RMSE to assess the goodness of our fits.

If space permits, we will selectively also provide the maximum error E_{max}

$$E_{max} = \max_i |F(X_i) - Y_i| \tag{3.42}$$

and the maximum relative error e_{max}

$$e_{max} = \max_i \frac{|F(X_i) - Y_i|}{|Y_i|} \tag{3.43}$$

of our fits to provide additional insight into their accuracy.

Methodology

4

In the following, we discuss how we apply the theory discussed in the previous chapter to study the relation between the tidal deformability and f-mode frequency for single neutron stars, as well as derive new universal relations for binary neutron star (BNS) mergers.

4.1 Data Set of Rotating Neutron Stars

Our own computations for this thesis only involve the tidal deformabilities of non-rotating neutron stars. For the f-mode frequencies, we rely on the data set of rotating neutron stars on which the results by Krüger and Kokkotas [KK20b, KK20a] are based: they use the `rns` code [SF95, NSGE98] to create the rotating equilibrium models for the same EoSs that we also consider in this thesis. They then developed their own perturbation code to compute the f-mode frequencies of the rotating neutron stars.

To analyze the dependence of the f-mode frequency on rotation, they consider three different types of sequences: for a constant value of either the central energy density ϵ_c, the gravitational mass M_0 or the baryon mass M_b, they vary the rotation rate of the star from irrotational to the Keplerian mass-shedding limit (at which the neutron star experiences instabilities).

They choose for each of these three sequence parameters a range of values to cover neutron stars from a gravitational mass of around 1 $M_\odot$, up to the maximum mass supported under uniform rotation, which, depending on the EoS, reaches values around $2.3-3.3M_\odot$ (the actual data does not quite reach these maximum values due to the stability of the perturbation code used to compute the f-mode frequencies). As such, they cover a wide range of neutron star configurations with varying rotation rates.

P. Manoharan, *Universal Relations for Binary Neutron Star Mergers with Long-lived Remnants*, BestMasters, https://doi.org/10.1007/978-3-658-36841-8_4

We rely on this data for not only the f-mode frequencies of these neutron stars, for both the co- and counter-rotating modes, but also other bulk properties such as the moment of inertia and effective compactness.

4.2 Relating Tidal Deformability to f-mode Frequency

While our primary focus for this thesis lies on the universal relations for binary neutron star mergers, our computations also allow, as a byproduct, for the investigation of universal relations for single neutron stars. After first revisiting previously proposed universal relations for single neutron stars, we will in particular consider the relation between the tidal deformability of a non-rotating star, and the f-mode frequency of the same star under varying degrees of rotation by following the sequences of rotating stars contained in the data set described above.

Both, the tidal deformability as well as the f-mode frequency play a central role in universal relations for neutron stars, and as such understanding their relation can shed light on the dynamics of neutron stars [CSLL14]. However, as we currently do not have a generally agreed upon formalism for the tidal deformability of rotating neutron stars [PGMF18], we investigate whether a meaningful relation between the non-rotating tidal deformability and the rotating f-mode can be found.

4.3 Binary Neutron Star Mergers

For our attempts of developing universal relations for BNS mergers, we propose a simplified model: we consider a binary of irrotational neutron stars with masses M_1 and M_2, and mass ratio $q = \frac{M_1}{M_2} \leq 1$. The merger of these two stars results in a long-lived remnant of mass $M = M_1 + M_2$ as we assume the mass loss during the merger to be negligible: numerical relativity simulations have shown that this mass loss is typically small, at the order of $10^{-4} - 10^{-2} M_\odot$ [BGJ13, HKK+13, SKK+16, RPH+18].

Furthermore, we assume the long-lived remnant to be a massive neutron star with mass $M \geq 2M_\odot$, up to the maximum mass for which the data set on rotating neutron stars provides f-mode frequencies.

To study the dependence of our relations on the rotation rate of the remnant, we allow angular rotation rates Ω between 0 (non-rotating) and the Kepler mass-shedding limit Ω_K (maximally rotating). However, we will below discuss constraints on Ω in an attempt to improve the accuracy of our relations under physical constraints.

For a wide range of the three primary parameters M, q and Ω, and a choice of EoS from the ones used in this thesis (cf. Section 3.1), we then compute the binary tidal deformability $\tilde{\Lambda}$ of the pre-merger objects, and compare it to the f-mode frequency and effective compactness of the remnant.

Implicitly, we assume that the long-lived remnant shares the same *cold* EoS as the pre-merger neutron stars and that the initially differential rotation of the remnant has already been driven to uniform rotation by viscous processes. Numerical relativity simulations show that these conditions are usually achieved at a cooling timescale of $\sim 2 - 3$s [RPBZ18].

We also do not consider the possibility of a phase-transition between the pre-merger and remnant EoS [BBB+19], i.e. the EoS remains the same throughout the merger. As discussed by Nandi and Pal [NP20], phase-transitions from purely hadronic matter to quark matter can be useful to address issues with reconciling tidal deformability upper bounds derived from the observation of `GW170817` [AoLV17] (implying a stiffer EoS), with the maximum mass lower bound derived from the observation of the millisecond pulsar `J0740+6620` [CFR+20] (implying a softer EoS).

Such phase-transitions, however, change the properties of neutron stars. In particular, they lead to, generally, smaller tidal deformabilities [HS19] compared to the purely hadronic case. As shown in [NP20] using the well-known I-Love-Q relation [YY17], this also affects the accuracy of universal relations that rely on the tidal deformability. As such, extending our work to include EoSs with phase-transitions will be an important direction for future work.

4.3.1 Post-merger Remnant

As described above, we assume the long-lived remnant of the merger to be a uniformly rotating, massive neutron star with mass $M \geq 2M_\odot$ and angular rotation rate Ω, or rotation frequency $\bar{\Omega} = \frac{\Omega}{2\pi}$. For the stellar parameters of these remnants, we utilize the dataset described in Section 4.1.

From this data set, we are able to take the frequency σ of the co-rotating ($l = 2$, $m = -2$) and counter-rotating ($l = 2$, $m = 2$) f-mode, and the effective compactness η of neutron stars across a wide range of masses and rotation rates, allowing for a comprehensive analysis of the relation between pre-merger tidal deformability and remnant properties.

While our initial results are general, and depend on the mass ratio of the pre-merger stars and rotation rate of the remnant, we also consider constraints on these parameters to further improve the accuracy of our results. To this end, we consider the universal relation for the rotation period of the long-lived remnant put forward by Radice et al. [RPBZ18]

$$P = \left[a \left(\frac{M_b}{M_\odot} - 2.5 \right) + b \right] \text{ms} \tag{4.1}$$

where a and b are EoS specific coefficients, and M_b the baryon mass of the long-lived remnant.

Results

5

In this chapter, we discuss the results of our analyses. We start by revisiting some of the existing universal relations for single neutron stars that we discussed in Chapter 2, before we turn towards the novel results we developed in the context of this thesis: first, we relate the tidal deformability of a non-rotating star to the f-mode frequency of a rotating counterpart. While these relations are not that interesting for practical purposes (since the neutron stars we might observe will either be rotating or non-rotating), our analysis produces some nice relations that can be of academic interest (as discussed in Section 4.2). Thereafter, we dive into our analyses of binary neutron star mergers with long-lived neutron star remnants, and present our new universal relations between the pre-merger stars' binary tidal deformability and the remnant's effective compactness.

5.1 Revisiting Existing Universal Relations

We discussed a number of universal relations for single neutron stars in Chapter 2 that relate their f-mode frequency σ, or their tidal deformability λ, to bulk properties of the star. We revisit them here to validate their accuracy through independent calculations and check their universality within the piecewise-polytropic EoSs we consider in this thesis. Note that all relations discussed here only consider non-rotating neutron stars.

Tsui and Leung [TL05b] relate the normalized, angular f-mode frequency $M_0\omega$ to the compactness C of a non-rotating neutron star through

Supplementary Information The online version contains supplementary material available at https://doi.org/10.1007/978-3-658-36841-8_5.

P. Manoharan, *Universal Relations for Binary Neutron Star Mergers with Long-lived Remnants*, BestMasters, https://doi.org/10.1007/978-3-658-36841-8_5

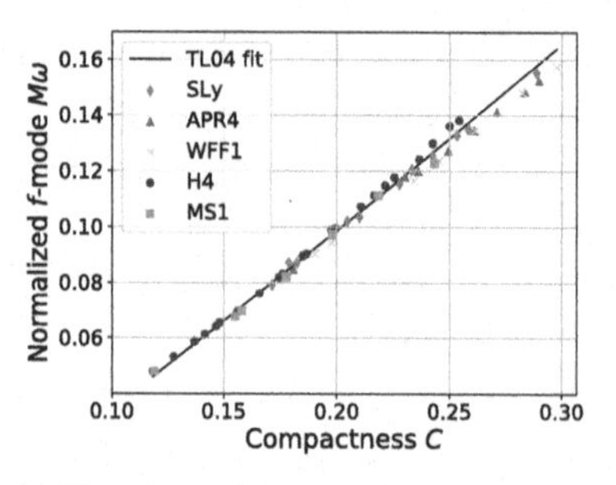

(a) The relation between the compactness and f-mode frequency by Tsui and Leung [TL05b].

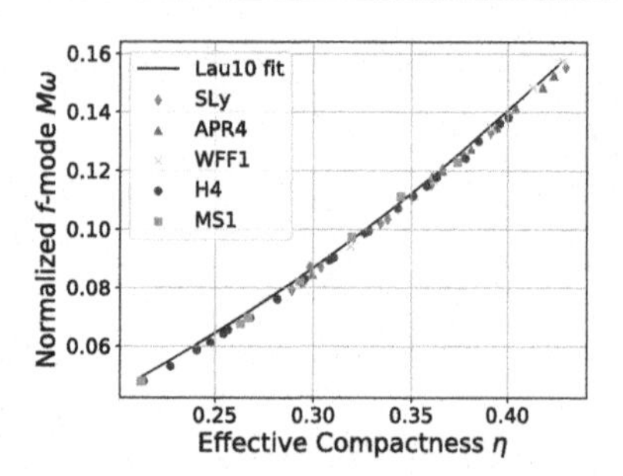

(b) The relation between the effective compactness and normalized f-mode frequency by Lau et al. [LLL10].

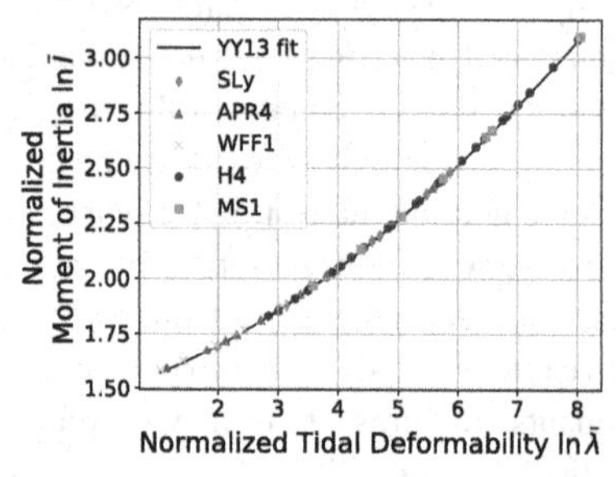

(c) The relation between the tidal deformability and the moment of inertia by Yagi and Yunes [YY13].

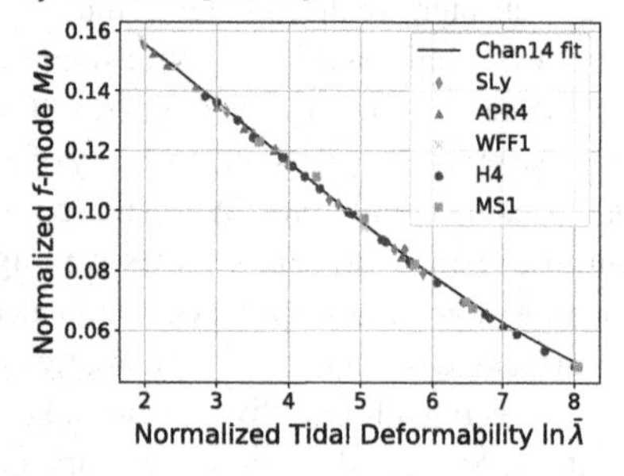

(d) The relation between the tidal deformability and normalized f-mode frequency by Chan et al. [CSLL14].

Figure 5.1 Previously known universal relations for single neutron tars with our own data

$$M_0\omega = aC^2 + bC + c. \tag{5.1}$$

where a, b and c are fitting constants. We compare their best fit for this relation with our own data in Figure 5.1a. We observe that our own data are fairly consistent with the relation put forward by Tsui and Leung. For high compactnesses, however, slight deviations from their fit can be observed, which was also observed by Chirenti et al. [CdK15] previously.

Lau et al. [LLL10] seek to improve upon the previous relations by instead relating the normalized f-mode to the effective compactness η through a relation of the form

$$M_0\sigma = a\eta^2 + b\eta + c \tag{5.2}$$

where a, b and c are again fitting constants. We present the comparison between their best fit for this relation to our own data in Figure 5.1b. We again observe a very strong correlation, however our own data consistently shows a slightly lower normalized f-mode than predicted by the relation in Equation (5.2). This can be explained by the different set of EoSs considered here than in [LLL10].

As for relations involving the tidal deformability, we first consider the well-known I-Love-Q relation put forward by Yagi and Yunes [YY13]. They propose a set of relations of the form

$$\ln y_i = a_i x_o + b_i \ln x_i + c_i (\ln x_i)^2 + d_i (\ln x_i)^3 + e_i (\ln x_i)^4 \tag{5.3}$$

that each involve two of the three eponymous quantities, respectively, with a_i, b_i, c_i, d_i and e_i being the fitting constants of the ith relation. We here exemplarily consider the one between the *normalized moment of inertia*

$$\bar{I} = \frac{I}{M_0^3} \tag{5.4}$$

and the *normalized tidal deformability*

$$\bar{\lambda} = \frac{\lambda}{M_0^5} \tag{5.5}$$

We illustrate the best fit presented in [YY13] together with our data in Figure 5.1c. We observe a very good fit of the original relation with our data points, further validating the accuracy and universality of this specific I-Love-Q relation for non-rotating stars.

Chan et al. [CSLL14] investigate relations between the f-mode frequency and the tidal deformability of non-rotating neutron stars, both important quantities in the other universal relations that we have discussed as they can be accurately measured (and/or constrained) through gravitational wave measurements. They present general relations of the form

$$M_0 \omega_{l'} = a_0 + a_1 \ln \bar{\lambda}_l + a_2 \left(\ln \bar{\lambda}_l\right)^2 + a_3 \left(\ln \bar{\lambda}_l\right)^3 + a_4 \left(\ln \bar{\lambda}_l\right)^4 \tag{5.6}$$

for the l-th multipole, normalized tidal deformability $\bar{\lambda}_l = \frac{\lambda_l}{M^{2l+1}}$ and the l'-th multipole, angular f-mode frequency ω_l, with the a_i being the fitting constants. We here only consider the special case of $l = l' = 2$ (as we do throughout this thesis). A comparison of this relation and our data is shown in Figure 5.1d. We again observe a very good agreement between the fit and our data, validating the relation put forward in [CSLL14].

5.2 Universal Relations for Single Neutron Stars

We now consider a rotating neutron star for which we compare its f-mode frequency with its non-rotating tidal deformability. As we discussed in Section 3.4.1, the $l = |m| = 2$ f-mode frequency of a rotating star splits into two different modes: the stable, co-rotating mode (for which we denote the frequency with σ^s), and the potentially unstable, counter-rotating mode (which we denote with σ^u). We investigate the relation of both these modes to the tidal deformability λ of the non-rotating star associated with the rotating star.

One question that arises in this context is how exactly we associate a rotating neutron star to its non-rotating counterpart. As we discussed in Section 4.1, the data set we use for the rotating neutron stars already provides sequences of rotating neutron stars of constant central energy density ϵ_c, gravitational mass M_0 or baryon mass M_b. For a given non-rotating star that fixes the value of one of these quantities, we thus consider a sequence of rotating neutron stars of increasing rotation rate that keep this quantity constant.

We are also free in how we represent the rotation rate of the rotating neutron stars. Here, we consider three different options for our analysis: first, the simple, angular rotation rate Ω. Second, the mass-normalized angular rotation rate $\hat{\Omega} = M_0\Omega$. And finally, third, the normalized angular rotation rate $\frac{\Omega}{\Omega_K}$ where Ω_K is the Keplerian (mass-shedding) rotation limit, i.e. the maximum possible rotation rate of the neutron star before it becomes unstable.

We now attempt, for both the co- and counter-rotating f-modes, to find, for each sequence type (with constant M_0, M_b or ϵ_c) separately, as well as all of them combined, a linear, parameterized relation of the form

$$\hat{\sigma} = M_0\sigma = a(x)\ln\bar{\lambda} + b(x) \tag{5.7}$$

where x is one the three rotational parameters discussed above. In our computations, a good fit is produced by quadratic polynomials for the coefficients $m(x)$ and $b(x)$, i.e.

$$a(x) = a_2x^2 + a_1x + a_0 \tag{5.8}$$

and

$$b(x) = b_2x^2 + b_1x + b_0 \tag{5.9}$$

We first discuss the combined fit for all sequences, which we illustrate in Figure 5.2. In these figures, we present a scatter plot of the tidal deformability against the co- or counter-rotating f-mode frequency at different rotation rates $\hat{\Omega} = \frac{M_0}{M_\odot}\frac{\Omega}{\mathrm{kHz}}$, for all sequence types. We can clearly observe a separation of the points by rotation

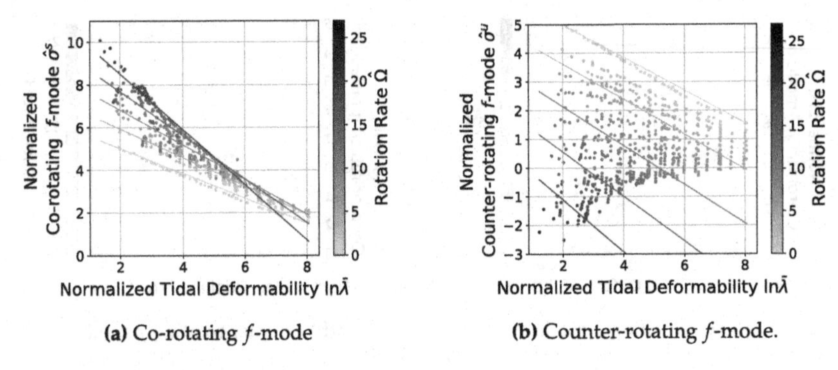

Figure 5.2 The $\hat{\Omega}$ dependent fits for the relation between the f-mode frequency and the non-rotating tidal deformability, with all sequence types

rate, which informed our choice of the fit proposed in Equation (5.7). In these plots, we also show the best linear fit to our data, at different rotation rate values $\hat{\Omega} \in \{0, 5, 10, 15, 20\}$ $M_{\odot}$k Hz, and we can observe a reasonably good fit.

Similar plots for each sequence type separately are shown in Figure 5.3 for the constant central energy density sequence, in Figure 5.4 for the constant gravitational mass sequence, and in Figure 5.5 for the constant baryon mass sequence. While the constant central energy density sequence shows very similar behavior to the combined sequence discussed above, the gravitational mass and baryon mass sequence show deviations, in particular for the co-rotating f-mode, where we either

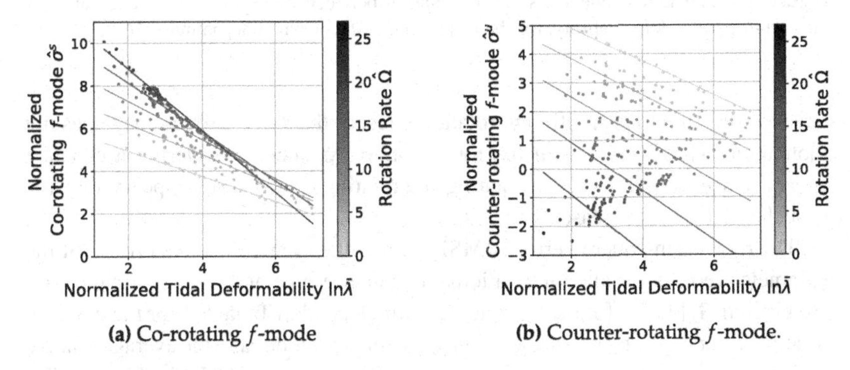

Figure 5.3 The $\hat{\Omega}$ dependent fits for the relation between the f-mode frequency and the non-rotating tidal deformability, for the sequences with constant energy density ϵ_c

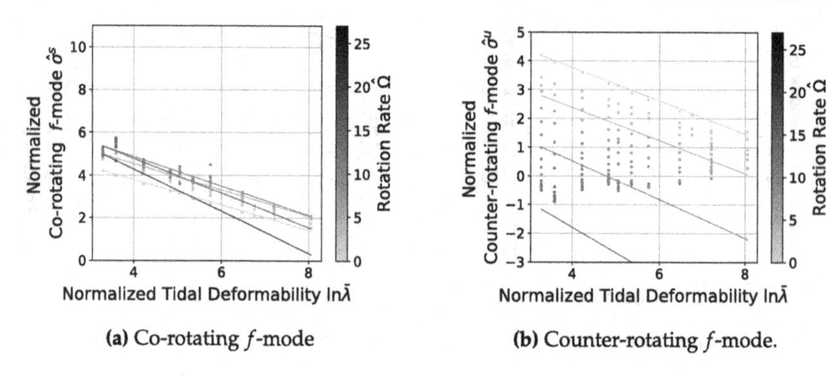

(a) Co-rotating f-mode

(b) Counter-rotating f-mode.

Figure 5.4 The $\hat{\Omega}$ dependent fits for the relation between the f-mode frequency and the non-rotating tidal deformability, for the sequences with constant gravitational mass M

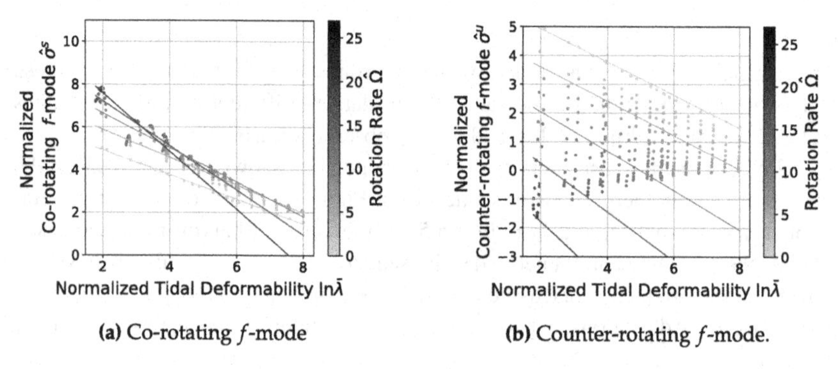

(a) Co-rotating f-mode

(b) Counter-rotating f-mode.

Figure 5.5 The $\hat{\Omega}$ dependent fits for the relation between the f-mode frequency and the non-rotating tidal deformability, for the sequences with constant baryon mass M_b

have overlap of points at different rotation rates, or the f-mode frequency does not monotonically increased with the rotation rate. As such, considering a constant energy density to compare non-rotating and rotating neutron stars appears to be the best choice for our use case.

The root-mean-squared error (RMSE) $\bar{E}$ and the average relative error $\bar{e}$ of the parameterized fits for all combinations of sequence type and rotational parameter are given in Table 5.1 for the co-rotating f-mode, and in Table 5.2 for the counter-rotating f-mode. Note that, in the counter-rotating f-mode case, the average relative error will typically be much larger we have frequencies around 0 Hz in this case. For

Table 5.1 Errors of the parameterized fits for the co-rotating f-mode for each rotational parameter—sequence types combination

$\bar{E}$	ρ_c	M	M_b	all
Ω	0.115	0.214	0.210	0.309
$M\Omega$	0.098	0.226	0.217	0.301
$\frac{\Omega}{\Omega_K}$	0.180	0.218	0.278	0.439

(a) Root-mean-squared error.

$\bar{e}$	ρ_c	M	M_b	all
Ω	0.015	0.037	0.029	0.042
$M\Omega$	0.010	0.037	0.030	0.042
$\frac{\Omega}{\Omega_K}$	0.023	0.037	0.044	0.066

(b) Average relative error.

Table 5.2 Errors of the parameterized fits for the counter-rotating f-mode for each rotational parameter—sequence types combination

$\bar{E}$	ρ_c	M	M_b	all
Ω	0.356	0.177	0.336	0.388
$M\Omega$	0.073	0.172	0.206	0.258
$\frac{\Omega}{\Omega_K}$	0.194	0.128	0.149	0.258

(a) Root-mean-squared error.

$\bar{e}$	ρ_c	M	M_b	all
Ω	0.397	0.435	0.937	0.782
$M\Omega$	0.074	0.356	0.551	0.503
$\frac{\Omega}{\Omega_K}$	0.192	0.297	0.457	0.568

(b) Average relative error.

a comparison between the co- and counter-rotating cases, the RMSE will therefore be much more useful.

As we can see, the RMSE is minimized by the combination of the sequences with constant constant energy density with the rotational parameter $\hat{\Omega}$, which thus presents the best fit for our use case. The exact coefficients for this fit are listed in Table 5.3 . The plots for the remaining sequence types, as well as the coefficients for the best fits, can be found in Appendix A.1 in the electronic supplementary material.

As we discussed in Section 4.2, the results presented here are mostly of academic interest: if we ever do observe single, rotating neutron stars through GW observations, we would measure, both, the rotating tidal deformability and rotating f-mode frequencies.

However, we will next discuss universal relations for BNS mergers that also relate the tidal deformability of non-rotating bodies (through the binary tidal deformability of the BNS) with the f-mode frequency of a rapidly rotating neutron star. The relation presented above thus gives a hint that the BNS relations do not appear by chance, but are a result of a general universal relation between tidal deformability

Table 5.3 Coefficients of the relation in Equation (5.7), for both, the co- and counter-rotating f-mode, considering only the sequences with constant energy density ϵ_c with $\hat{\Omega}$ as the rotational parameters. The coefficients are defined as in Equations (5.8) and (5.9)

i	a_i	b_i
2	-1.168×10^{-3}	-1.761×10^{-3}
1	-1.932×10^{-2}	3.132×10^{-1}
0	-5.968×10^{-1}	6.130

(a) For the co-rotating f-mode.

i	a_i	b_i
2	-6.976×10^{-4}	-3.663×10^{-3}
1	-7.421×10^{-3}	-1.747×10^{-1}
0	-5.888×10^{-1}	6.095

(b) For the counter-rotating f-mode.

and f-mode frequency for neutron stars that survives a separation by rotation, and that warrants further investigation in future work.

5.3 f-mode Relations for BNS Mergers

We now consider the BNS case and try to relate the binary tidal deformability of a BNS to the f-mode frequencies of the long-lived remnant. We will begin by considering the co-rotating f-mode, as that is what corresponds to the f-mode frequency considered by Kiuchi et al. [KKK+20], and we will try to produce a similar relation to the one proposed there.

Afterwards, we will try to produce a universal relation for the counter-rotating f-mode as well.

5.3.1 Co-rotating f-mode

To find a relation similar to the one presented by Kiuchi et al. [KKK+20], we compute the binary tidal deformability $\tilde{\Lambda}$ of a BNS characterized by a total mass M and a mass ratio q, and compare it to the *normalized, co-rotating f-mode frequency*

$$\hat{\sigma}^s = \frac{M}{M_\odot}\frac{\sigma^s}{\text{k Hz}} \tag{5.10}$$

of the long-lived remnant at different angular rotation rates Ω, as computed in [KK20b]. Note that we now use M to denote the total mass of the binary neutron star, in contrast M_0, which we used to denote the mass of a single neutron star.

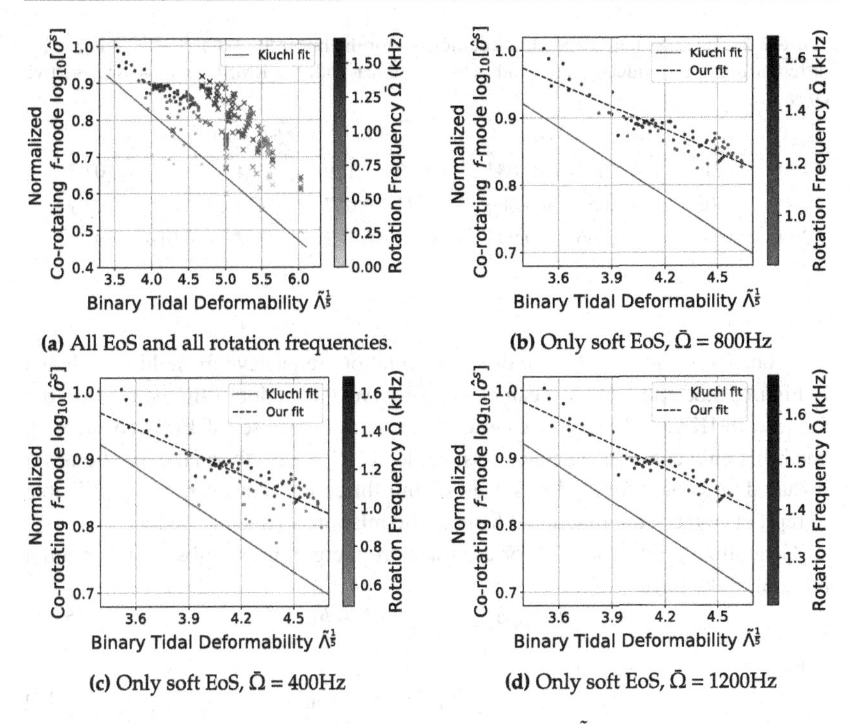

(a) All EoS and all rotation frequencies.

(b) Only soft EoS, $\bar{\Omega} = 800$Hz

(c) Only soft EoS, $\bar{\Omega} = 400$Hz

(d) Only soft EoS, $\bar{\Omega} = 1200$Hz

Figure 5.6 The relation between the binary tidal deformability $\tilde{\Lambda}$ and the co-rotating f-mode frequency σ for BNS with mass ratio $q = 1$

In Figure 5.6a, we show an initial plot of all data points together with the fit proposed in [KKK+20]. We separate the soft EoSs (dots) from the stiff EoSs (crosses) (cf. Section 3.1) by marker style. As we can see, without further constraints, we do not immediately obtain a linear relation as proposed by Kiuchi et al. [KKK+20]. In fact, we observe a clear distinction between the soft and stiff EoSs, and stars with low rotation frequencies also do not seem to fit into a potential linear relation.

To remedy these issues, we add two constraints: first, since the EoSs used in [KKK+20] correspond mostly to the soft EoSs considered here, we, for now, ignore the stiff EoSs. And second, as we expect the remnant to be fast rotating, we require it to have a rotation frequency $\bar{\Omega} = \Omega/2\pi$ above a minimum rotation frequency threshold $\bar{\Omega}_{thr} \in \{400, 800, 1200\}$Hz.

The resulting relation that includes the linear fit for stars with soft EoSs (i.e. APR4, SLy and WFF1) and rotation frequencies $\bar{\Omega} \geq 800$Hz is shown in Fig-

Table 5.4 The mass ratio dependent coefficients for the linear fit of the $\tilde{\Lambda} - \hat{\sigma}^s$ relation for different rotation frequency thresholds $\bar{\Omega}_{thr}$, together with the RMSE and average relative error

$\bar{\Omega}_{th}$[Hz]	a_2	a_1	a_0	b_2	b_1	b_0	$\bar{E}$	$\bar{e}$
400	0.033	−0.065	−0.084	−0.060	0.119	1.304	0.024	0.020
800	0.029	−0.058	−0.086	−0.044	0.087	1.322	0.015	0.014
1200	0.034	−0.066	−0.093	−0.052	0.102	1.358	0.014	0.012

ure 5.6b. The same plots for the two other rotation frequency thresholds are shown in Figures 5.6c and 5.6d. We can clearly observe a deviation from the relation put forward in [KKK+20] that is caused by a) the different set of EoSs considered, and b) us only considering cold EoSs, while in [KKK+20] thermal corrections are included to account for hot EoSs. Despite this difference, however, we are still able to reproduce the same functional form of the relation between $\tilde{\Lambda}$ and $\hat{\sigma}$.

Generalizing this linear fit for arbitrary pre-merger mass ratios q, we obtain a relation of the form

$$\log_{10} \hat{\sigma} = a(q) \cdot \tilde{\Lambda}^{\frac{1}{5}} + b(q) \tag{5.11}$$

where

$$a(q) = a_2 q^2 + a_1 q + a_0 \tag{5.12}$$

and

$$b(q) = b_2 q^2 + b_1 q + b_0 \tag{5.13}$$

The exact values of these coefficients for different rotation frequency thresholds are given in Table 5.4. We also give the RMSE $\bar{E}$ and the average relative error $\bar{e}$ of the estimated values for $\hat{\sigma}$ through Equation (5.11) compared to the original value (only considering stars with rotation frequency $\bar{\Omega} \geq \bar{\Omega}_{thr}$). As one would expect, the higher the rotation frequency threshold is chosen, the more accurate the linear fit becomes. For instance, with $\bar{\Omega}_{thr} = 800$Hz, our linear fit achieves an RMSE of 0.015, and an average relative error of $\sim 1.4\%$.

5.3.2 Counter-rotating f-mode

We now try to relate the binary tidal deformability to the *normalized, counter-rotating f-mode frequency*

$$\hat{\sigma}^u = \frac{M}{M_\odot} \frac{\sigma^u}{\text{k Hz}} \tag{5.14}$$

of the long-lived remnant, again at different rotation frequencies $\bar{\Omega}$ and for different mass ratios q. We also continue to only consider the soft EoSs (APR4, WFF1 and SLy), as before, but will later try to relax this requirement.

Since the counter-rotating f-mode frequency can reach negative values, we cannot straightforwardly adapt the same form for our relation as above due to the logarithm used there. While there are several approaches one could take to alleviate this issue (e.g. ignore negative frequencies, or only consider the absolute value of the frequency), our evaluations have shown that the best results are achieved when one attempts a simple linear relation.

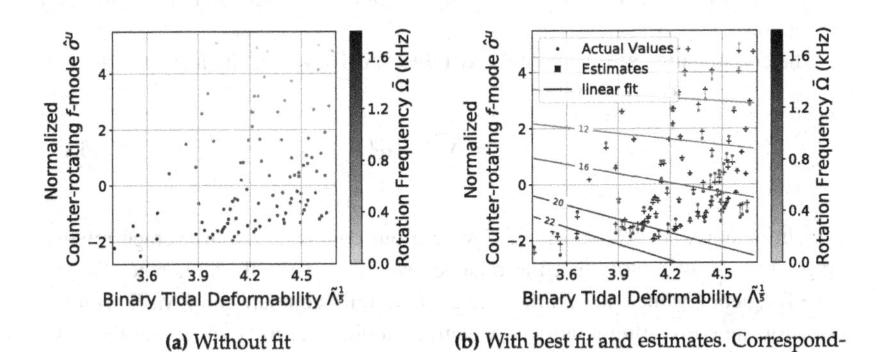

(a) Without fit

(b) With best fit and estimates. Corresponding pairs are connected by a line.

Figure 5.7 Counter-rotating f-mode frequency against the binary tidal deformability, with and without the rotation dependent fit, for $q = 1$

In Figure 5.7a, we show the counter-rotating f-mode against the binary tidal deformability, colored by the rotation frequency of the star. We can immediately see that a single, rotation independent relation is clearly out of reach. However, we do observe a clear separation of the points by said rotation frequency, suggesting a linear relation of the form

$$\hat{\sigma}^u = a(q, \hat{\Omega})\tilde{\Lambda}^{\frac{1}{5}} + b(q, \hat{\Omega}) \tag{5.15}$$

where

$$\hat{\Omega} = \frac{M}{M_\odot} \frac{\Omega}{\text{k Hz}} \tag{5.16}$$

is the *normalized angular rotation rate*, and a and b are given by

$$a(q, \hat{\Omega}) = a_2(q)\hat{\Omega}^2 + a_1(q)\hat{\Omega} + a_0(q) \tag{5.17}$$

$$b(q, \hat{\Omega}) = b_2(q)\hat{\Omega}^2 + b_1(q)\hat{\Omega} + b_0(q) \tag{5.18}$$

where each of the q-dependent coefficients is again a quadratic function of q. While we experimented with different parameterizations for the rotation rate, the normalized angular rotation rate $\hat{\Omega}$ again proved to provide the best fit. For the sake of brevity, we will, in the following, only present the results using this parameterization.

Our best fit for this relation is illustrated in Figure 5.7b for $\hat{\Omega} \in \{2.5, 7.5, 12, 16, 20, 22\}$ $M_\odot$k Hz . We also show our estimates for each of the data points, indicated by squares.

The coefficients for this fit are listed Table 5.5. They have the form

$$x_i = \sum_{j=0}^{2} x_{i,j} q^j \tag{5.19}$$

This fit achieves an RMSE of 0.183 within our data set, and an average relative error of $\sim$ 18.4%. Note that the relative error is necessarily higher here as we have frequency values around the origin. If we only compare the RMSE with the error for the co-rotating relation, we achieve similar accuracy (note that there we estimated the logarithm of the normalized f-mode frequency). However, we now have a 2-parameter fit, which is more difficult to utilize in practice, as in addition to an estimate for the mass ratio, we would also need an accurate estimate for the rotation frequency of the remnant.

Table 5.5 Coefficients for the relation in Equation (5.15), considering only soft EoSs

j	2	1	0
$a_{2,j}$	6.875×10^{-4}	-1.330×10^{-3}	-3.529×10^{-3}
$a_{1,j}$	1.364×10^{-2}	-2.721×10^{-2}	2.536×10^{-2}
$a_{0,j}$	4.560×10^{-3}	-9.219×10^{-3}	2.357×10^{-1}
$b_{2,j}$	1.430×10^{-3}	-2.889×10^{-3}	1.147×10^{-2}
$b_{1,j}$	-9.279×10^{-2}	1.839×10^{-1}	-3.204×10^{-1}
$b_{0,j}$	1.950×10^{-1}	-3.809×10^{-1}	6.048

Still, in cases where we are able to obtain accurate estimates for these two quantities for a BNS merger, we can use the relation presented here to predict whether we might observe, e.g., a CFS-instability in the remnant due to a negative counter-rotating f-mode frequency [Cha70, FS78].

Reintroducing stiff EoS

In contrast to the case with the co-rotating f-mode frequency, the 2-parameter fit we developed for the counter-rotating f-mode might allow us to reintroduce the stiff EoSs (MS1 and H4) without losing too much of our accuracy. The plots resulting from reintroducing the stiff EoSs are shown in Figure 5.8. We again indicate the data points corresponding to the stiff EoSs with crosses. While we can see that the stiff EoS solutions follow a similar trajectory as the soft EoSs, trying to fit a similar function as in Equation (5.15) to this data does not produce as good of a fit: the best fit achieves an RMSE of 0.3688, and an average relative error of $\sim 41.8\%$. Again, the relative error is emphasized due to the value close to zero. But overall, we still approximately double the error by including the stiff EoSs as well.

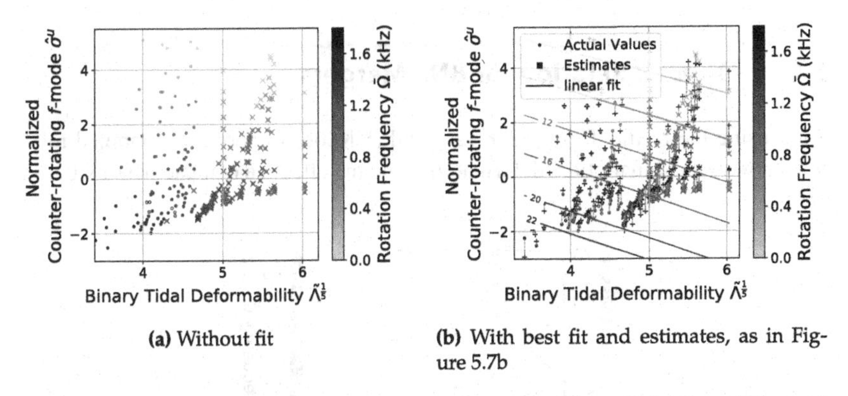

(a) Without fit

(b) With best fit and estimates, as in Figure 5.7b

Figure 5.8 Counter-rotating f-mode frequency against the binary tidal deformability, with and without the rotation dependent fit, for $q = 1$. Including soft and stiff EoSs

We also attempted using other, higher order fits between the normalized f-mode frequency and the binary tidal deformability. However, none showed any meaningful improvement over the results presented above. As such this relation seems to be only useful if we consider soft EoSs.

5.4 Combined η Relation for BNS Mergers

Combining the linear $\tilde{\Lambda} - \hat{\sigma}^s$ fit in Equation (5.11) with the $\eta - \hat{\sigma}^s$ fit in [KK20b, Equation (6)], we obtain a combined $\tilde{\Lambda} - \eta$ relation of the form

$$\eta = \frac{10^{a(q)\cdot\tilde{\Lambda}^{\frac{1}{5}}+b(q)} - \left(c_1 + c_2\hat{\Omega} + c_3\hat{\Omega}^2\right)}{d_1 + d_3\hat{\Omega}} \tag{5.20}$$

where the coefficients c_1, c_2, c_3, d_1 and d_3 are given in [KK20b]. This relation allows us to estimate the effective compactness η of the post-merger star from the tidal deformabilities of the pre-merger neutron stars.

A comparison of the estimated effective compactnesses through Equation (5.20) and the actual effective compactnesses is illustrated in Figures 5.9 and 5.10.

The RMSE, average relative error and the maximum relative error for each rotation frequency threshold $\bar{\Omega}_{thr}$ are given in Table 5.6. On average, we achieve a relative error $\bar{e}$ of around $2\% - 4\%$.

5.5 Direct η Relation for BNS Mergers

After using the results from [KKK+20] and [KK20b] to derive a combined relation between the binary tidal deformability and the effective compactness by going

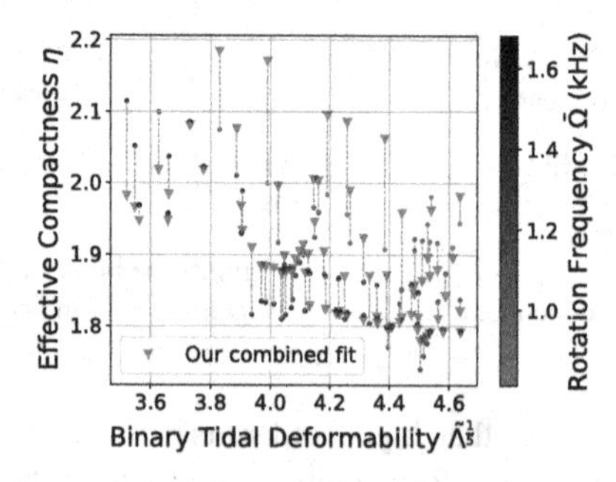

Figure 5.9 Comparison of the estimated effective compactnesses through Equation (5.20) (triangles), and the actual effective compactnesses (dots) for $q = 1$ and $\bar{\Omega}_{thr} = 800$Hz. Corresponding pairs are connected by a line

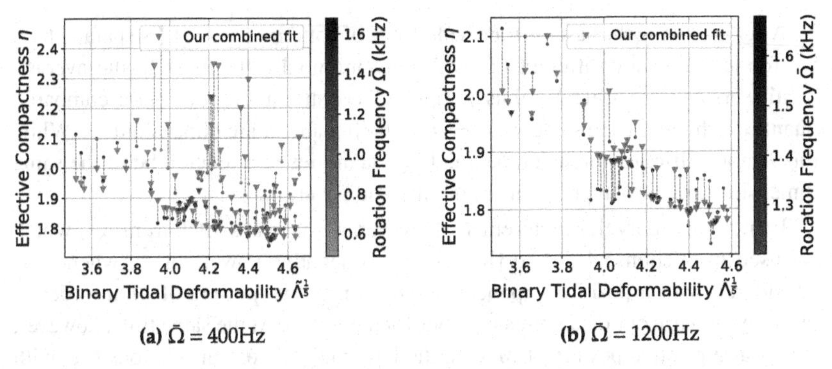

Figure 5.10 Same as Figure 5.9, but for $\bar{\Omega}_{thr} = 400$Hz and $\bar{\Omega}_{thr} = 1200$Hz, for $q = 1$

through the f-mode frequency, we now attempt a direct relation between these two quantities with the goal of achieving improved accuracy.

5.5.1 Direction Relation with soft EoSs

As with the combined relation in Section 5.4, we begin by considering the EoSs that we call *stiff*, i.e. WFF1, APR4 and SLy (cf. Section 3.1). We compare the values of $\tilde{\Lambda}$ and η for the full range of mass rations and remnant rotation rates, introducing varying powers of the binary's total mass $\bar{M}$ to both sides until we find the best linear fit. The general form of the relation thus is

$$\log_{10}\left[\bar{M}^i \eta\right] = a(q, \hat{\Omega})\bar{M}^j \tilde{\Lambda}^k + b(q, \hat{\Omega}) \tag{5.21}$$

where i and j are integers, and k is a rational fraction.

Table 5.6 The RMSE $\overline{E}$, the average relative error $\bar{e}$, and the maximum relative error e_{max} of the combined fit (cf. Equation (5.20)) at different rotation frequency thresholds $\bar{\Omega}_{thr}$

$\bar{\Omega}_{thr}$[Hz]	400	800	1200
$\overline{E}$	0.1085	0.0583	0.0464
$\bar{e}$	0.0367	0.0240	0.0205
e_{max}	0.1970	0.0872	0.0580

After analyzing the best fits for a wide range of combinations of these parameters, we found that many different fits with only minimal differences in the average relative error are possible. We thus add another constraint: the units of the compared quantities should be the same, i.e. $i = j$, as both η and $\tilde{\Lambda}$ are dimensionless. While this is not a strict requirement, having both quantities in a relation be of the same dimension typically improves its reliability and accuracy.

In order to satisfy this requirement, however, we need to switch from $\tilde{\Lambda}^{\frac{1}{5}}$, which we used in Section 5.4 due to [KKK+20], to negative powers of $\tilde{\Lambda}$: when $\tilde{\Lambda}$ is considered with a positive exponent, we need negative powers of $\bar{M}$ in order to properly *squeeze* our data points together for a good fit. At the same time, however, we require positive powers of $\bar{M}$ with the left hand side of our relations (i.e. with η) to properly scale that part of the relation. Consequently, if we want to achieve the same dimensions on both sides of the relation, we have to consider the relation between η and $\tilde{\Lambda}^{-k}$, for some positive value k.

In the following we present, for the sake of comparison, the best result for $\tilde{\Lambda}^{-\frac{1}{5}}$, as the average relative error for the overall best combination of parameters is only marginally better (at the order of 10^{-3}). A comparison of the errors for some of the best parameter combinations is given in Appendix A.2 in the electronic supplementary material.

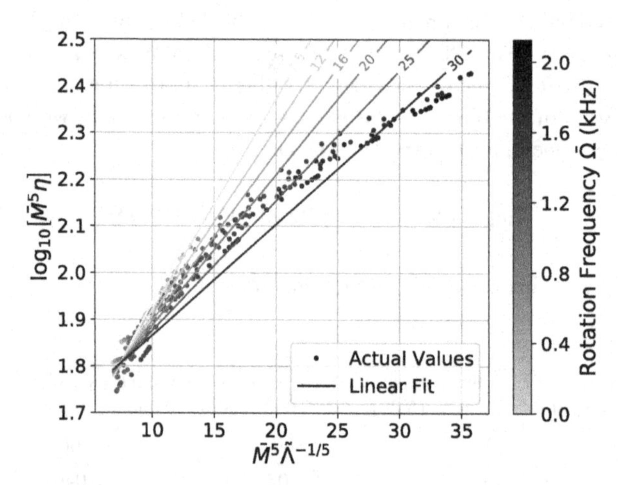

Figure 5.11 Our best fit for the relation in Equation (5.22), considering only soft EoS, for $q = 1$

Table 5.7 Coefficients of the direct relation in Equation (5.22), considering only soft EoSs

j	2	1	0
$a_{2,j}$	-2.993×10^{-6}	5.904×10^{-6}	-5.045×10^{-6}
$a_{1,j}$	1.037×10^{-4}	-2.054×10^{-4}	-1.145×10^{-3}
$a_{0,j}$	6.833×10^{-3}	-1.335×10^{-2}	6.053×10^{-2}
$b_{2,j}$	-6.101×10^{-5}	1.192×10^{-4}	-1.162×10^{-4}
$b_{1,j}$	-6.309×10^{-4}	1.256×10^{-3}	8.612×10^{-3}
$b_{0,j}$	1.087×10^{-2}	-2.149×10^{-2}	1.415

For the relation between $\bar{M}^i\eta$ and $\bar{M}^i\tilde{\Lambda}^{-\frac{1}{5}}$, we achieve the smallest average relative error for $i = 5$, i.e. the relation has the form

$$\log_{10}\left[\bar{M}^5\eta\right] = a(q,\hat{\Omega})\bar{M}^5\tilde{\Lambda}^{-\frac{1}{5}} + b(q,\hat{\Omega}) \tag{5.22}$$

where the coefficients are again defined as in Equations (5.17) and (5.18). The best fit of this relation to our data is illustrated in Figure 5.11 for $q = 1$. Over all mass ratios q, this fit achieves an RMSE of 0.02, an average relative error of $\bar{e} \sim 0.8\%$, and a maximum relative error of $e_{max} \sim 3.1\%$. The coefficients for this fit are listed in Table 5.7.

The direct relation thus shows improved accuracy compared to the combined relation presented earlier: we go from an average relative error of 2.4% to 1%. Furthermore, as we will see next, the direct relation also admits the stiff EoSs that we previously excluded for the combined relation, thus improving on universality compared to the combined relation presented earlier.

5.5.2 Reintroducing stiff EoSs

We next fit the relation in Equation (5.20) to the extended data set that contains stars of, both, soft and stiff EoSs (H4 and MS1). The resulting best fit is shown in Figure 5.12. Here, the data points for the stiff EoS are indicated by crosses. This new fit for all EoSs achieves an RMSE of 0.046, an average relative error of $\sim 1.7\%$, and a maximum relative error of $\sim 7.6\%$. The coefficients of this fit are listed in Table 5.8. As we can see, the direct relation also admits the stiff EoSs with close to double the error. However, the error remains small and we achieve high accuracy even including stiff EoSs.

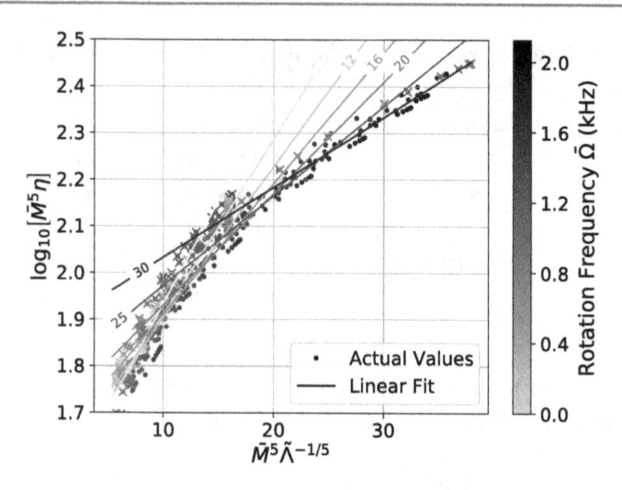

Figure 5.12 The same as in Figure 5.11, but with soft and stiff EoSs. Stiff EoSs are indicated by crosses

Table 5.8 Coefficients of the direct relation in Equation (5.22), considering soft and stiff EoSs

j	2	1	0
$a_{2,j}$	-2.442×10^{-6}	4.776×10^{-6}	7.235×10^{-6}
$a_{1,j}$	-1.153×10^{-5}	2.293×10^{-5}	-1.380×10^{-3}
$a_{0,j}$	4.895×10^{-3}	-9.585×10^{-3}	5.213×10^{-2}
$b_{2,j}$	7.236×10^{-5}	-1.4118×10^{-4}	3.512×10^{-4}
$b_{1,j}$	-1.894×10^{-3}	3.702×10^{-3}	3.216×10^{-3}
$b_{0,j}$	1.553×10^{-2}	-3.034×10^{-2}	1.494

5.6 Direct Relation with Constraints

The direct relation discussed in the previous Section allows for the full range of rotation frequencies of the long-lived remnant, from non-rotating up to the Keplerian limit. As such, imposing additional constraints to the remnant appears to be a straightforward way to further improve the accuracy of our relation.

Constraining the post-merger remnant's rotation rate according to Equation (4.1), we end up with rotation frequencies around $\bar{\Omega} \sim 950 - 1150$Hz (depending on the remnants baryon mass). Considering only neutron stars with rotation rates in a 10% range around this value produces the relation presented in Figure 5.13. This relation again allows for a linear fit, however this time only dependent on the mass ratio q, i.e. we have a relation of the form

$$\log_{10}\left[\bar{M}^5\eta\right] = a(q)\bar{M}^5\tilde{\Lambda}^{-\frac{1}{5}} + b(q) \tag{5.23}$$

where the coefficients a and b are only dependent on q. The coefficients have the form

$$x = \sum_{i=0}^{2} x_i q^i \tag{5.24}$$

and their values for the best fits are listed in Table 5.10.

A comparison of the achieved accuracy with the more general relation presented earlier, both for the cases of only considering soft EoSs as well as all EoSs, is presented in Table 5.9. While we can observe a divergence of the actual data points from our fit for the case with all EoSs (cf. Figure 5.13b), we see in Table 5.9 that

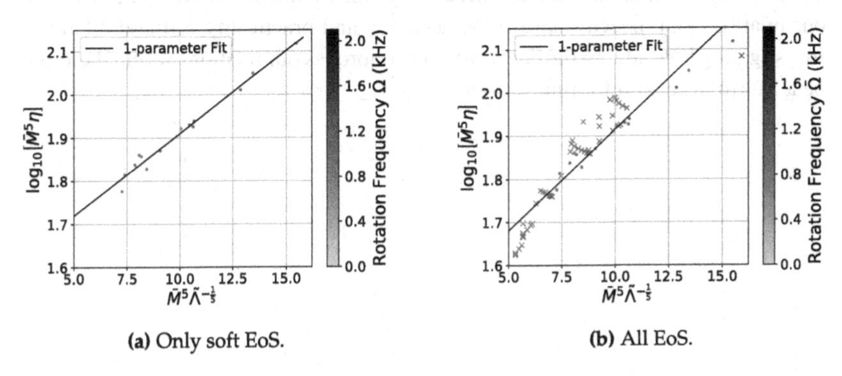

(a) Only soft EoS. (b) All EoS.

Figure 5.13 The relation in Equation (5.23) for $q = 1$, constrained to rotation rates according to Equation (4.1)

Table 5.9 Average relative error $\bar{e}$ and maximum relative error e_{max} achieved for the relation in Equations (5.22) (all Ω) and (5.23) (fixed Ω). Fixed Ω has rotation rates constrained by Eq. (4.1)

		$\overline{E}$	$\bar{e}$	e_{max}
stiff EoS	all Ω	0.021	0.008	0.031
	fixed Ω	0.014	0.006	0.017
all EoS	all Ω	0.046	0.017	0.076
	fixed Ω	0.042	0.017	0.052

Table 5.10 Coefficients of the constrained relation in Equation (5.23)

i	a_i	b_i
2	7.185×10^{-3}	-7.877×10^{-3}
1	-1.406×10^{-2}	1.544×10^{-2}
0	4.511×10^{-2}	1.520

(a) Only soft EoSs

i	a_i	b_i
2	7.451×10^{-3}	-4.176×10^{-3}
1	-1.459×10^{-2}	8.220×10^{-3}
0	5.391×10^{-2}	1.443

(b) For all EoSs

restricting ourselves to a specific rotation rate range indeed improves the accuracy of the fit slightly. In addition, this new constrained relation reduces the number of free parameters from previously two (q and $\hat{\Omega}$), to now only the mass ratio q, making it much more useful in practice: observations might, e.g., have difficulties with accurately estimating the rotation rate of the remnant star.

As such, we have shown that introducing constraints on the free parameters in our relations can indeed improve the accuracy and practicality of these relations. Investigating how adding further constraints improves our relations would therefore be a fruitful direction for future work.

Discussion & Conclusion 6

In this thesis, we proposed a novel approach to developing universal relations for BNS mergers using results from perturbative calculations instead of using full-fledged numerical relativity simulations.

Our approach is based on a simplified model for a BNS merger that results in a long-lived neutron star remnant: instead of performing a detailed simulation of the merger process, we individually consider the pre-merger phase, represented by two irrotational neutron stars, and the long-lived remnant, represented by a rapidly rotating neutron star. By comparing the neutron stars in these two phases, we then obtain relations connecting their properties.

Inspired by previous results by Kiuchi et al. [KKK+20], and using the results from [KK20b], we proposed a novel universal relation between the pre-merger binary tidal deformability $\tilde{\Lambda}$ and the effective compactness η of the post-merger remnant with high accuracy, reaching average relative error of $\sim 1\%$.

As the binary tidal deformability can be constrained fairly well from the observation of the GWs emitted during the pre-merger phase [FH08, Fav14], and as the spectrum of the pre-merger GW waves lie very well within the sensitivity range of current-generation GW detectors [RFR+14], the relations presented in this thesis can be useful for predicting the stellar parameters of a potential long-lived remnant produced by a BNS merger.

They can, in particular, be useful for further constraining the true EoS of neutron stars: for instance, as recently discussed by Greif et al. [GHL+20], combining accurate estimates for the mass M_0 (found directly from the gravitational waves) and the moment of inertia I (derived by using Equation (1.1) to estimate $\eta = \sqrt{M_0^3/I}$), can put further constraints on the radius of the neutron star, and thus its EoS.

The relations presented here might also become useful in the context of electromagnetic (EM) observations: Yamasaki et al. [YTK20b] recently proposed, under

P. Manoharan, *Universal Relations for Binary Neutron Star Mergers with Long-lived Remnants*, BestMasters, https://doi.org/10.1007/978-3-658-36841-8_6

the assumption that the fast radio burst `FRB 181112` originates from the remnant of a BNS merger, a novel universal relation between the spin period of the remnant (derived from the pulse timing of the FRB) and the binary tidal deformability of its pre-merger stars. Combining their relation with Equation (1.1), we obtain a universal relation between the effective compactness of the remnant and its spin period. While their assumptions on the origin of the FRB could be considered strenuous, the estimation of a pulsar spin period is still possible through traditional radio observations [LK12]. Our universal relation would then also be useful in conjunction with purely electromagnetic observations.

While we only used perturbative calculations for our analyses, they were not performed in a vacuum: numerical relativity simulations still play an important role in directing our efforts into the right direction. Not only were our efforts inspired by previous, simulation-based results by Kiuchi et al. [KKK+20], but also, by employing the universal relation for the remnant spin period put forward by Radice et al. [RPBZ18], we were able to further constrain our universal relations to more physical rotation rates of the remnant, improving our accuracy, and reducing the number of free parameters. As such, the approach presented in this thesis should mostly be considered supplementary to existing numerical relativity efforts.

6.1 Future Directions

As already mentioned in the introduction, our model for the BNS merger is markedly simple and does not cover all aspects and details relevant for the merger of two neutron stars. Our approach, however, can be freely extended to include more involved measures such hot EoSs to model the remnant right after the merger before it cools down. As discussed in Section 4.3, considering EoSs with phase-transitions, specifically, will be an important direction for future work.

Of interest would also be to include differential rotation into our post-merger model and to compute the corresponding oscillation modes (and other stellar parameters). This task has already been performed for simple differential rotation models under various approximations, such as the Cowling approximation [YRKE02, KGK10] and conformal-flatness [SAF04]. However, a more comprehensive approach with a more accurate model of differential rotation, and the treatment of the oscillations mod3es without approximation, would be desirable.

Bibliography

[AK98] Nils Andersson and Kostas D. Kokkotas. Towards gravitational wave asteroseismology. *MNRAS*, 299(4):1059–1068, October 1998.

[AKK96] Nils Andersson, Yasufumi Kojima, and Kostas D. Kokkotas. On the Oscillation Spectra of Ultracompact Stars: an Extensive Survey of Gravitational-Wave Modes. *ApJ*, 462:855, May 1996.

[AoLV17] B. P. Abbott, others, LIGO Scientific Collaboration, and Virgo Collaboration. GW170817: Observation of Gravitational Waves from a Binary Neutron Star Inspiral. *Phys. Rev. Lett.*, 119(16):161101, October 2017.

[AoLV20] B. P. Abbott, others, LIGO Scientific Collaboration, and Virgo Collaboration. GW190425: Observation of a Compact Binary Coalescence with Total Mass $\sim 3.4 M_{\odot}$. *ApJ*, 892(1):L3, March 2020.

[APR98] A. Akmal, V. R. Pandharipande, and D. G. Ravenhall. Equation of state of nucleon matter and neutron star structure. *Phys. Rev. C*, 58(3):1804–1828, September 1998.

[BBB+19] Andreas Bauswein, Niels-Uwe F. Bastian, David B. Blaschke, Katerina Chatziioannou, James A. Clark, Tobias Fischer, and Micaela Oertel. Identifying a First-Order Phase Transition in Neutron-Star Mergers through Gravitational Waves. *Phys. Rev. Lett.*, 122(6):061102, February 2019.

[BBF99] Omar Benhar, Emanuele Berti, and Valeria Ferrari. The imprint of the equation of state on the axial w-modes of oscillating neutron stars. *MNRAS*, 310(3):797–803, December 1999.

[BBV+20] Andreas Bauswein, Sebastian Blacker, Vimal Vijayan, Nikolaos Stergioulas, Katerina Chatziioannou, James A. Clark, Niels-Uwe F. Bastian, David B. Blaschke, Mateusz Cierniak, and Tobias Fischer. Equation of State Constraints from the Threshold Binary Mass for Prompt Collapse of Neutron Star Mergers. *Phys. Rev. Lett.*, 125(14):141103, October 2020.

[BDN15] Sebastiano Bernuzzi, Tim Dietrich, and Alessandro Nagar. Modeling the Complete Gravitational Wave Spectrum of Neutron Star Mergers. *Phys. Rev. Lett.*, 115(9):091101, August 2015.

[BFG04] Omar Benhar, Valeria Ferrari, and Leonardo Gualtieri. Gravitational wave asteroseismology reexamined. *Phys. Rev. D*, 70(12):124015, December 2004.

P. Manoharan, *Universal Relations for Binary Neutron Star Mergers with Long-lived Remnants*, BestMasters, https://doi.org/10.1007/978-3-658-36841-8

[BGJ13] A. Bauswein, S. Goriely, and H. T. Janka. Systematics of Dynamical Mass Ejection, Nucleosynthesis, and Radioactively Powered Electromagnetic Signals from Neutron-star Mergers. *ApJ*, 773(1):78, August 2013.

[BJ12] A. Bauswein and H. T. Janka. Measuring Neutron-Star Properties via Gravitational Waves from Neutron-Star Mergers. *Phys. Rev. Lett.*, 108(1):011101, January 2012.

[BJHS12] A. Bauswein, H. T. Janka, K. Hebeler, and A. Schwenk. Equation-of-state dependence of the gravitational-wave signal from the ring-down phase of neutron-star mergers. *Phys. Rev. D*, 86(6):063001, September 2012.

[BNB+14] Sebastiano Bernuzzi, Alessandro Nagar, Simone Balmelli, Tim Dietrich, and Maximiliano Ujevic. Quasiuniversal Properties of Neutron Star Mergers. *Phys. Rev. Lett.*, 112(20):201101, May 2014.

[BP09] Taylor Binnington and Eric Poisson. Relativistic theory of tidal Love numbers. *Phys. Rev. D*, 80(8):084018, October 2009.

[BR16] Cosima Breu and Luciano Rezzolla. Maximum mass, moment of inertia and compactness of relativistic stars. *MNRAS*, 459(1):646–656, June 2016.

[BSB18] Gabriele Bozzola, Nikolaos Stergioulas, and Andreas Bauswein. Universal relations for differentially rotating relativistic stars at the threshold to collapse. *MNRAS*, 474(3):3557–3564, March 2018.

[BWG+20] Ernesto Benitez, Joseph Weller, Victor Guedes, Cecilia Chirenti, and M. Coleman Miller. Investigating the I-Love-Q and w-mode Universal Relations Using Piecewise Polytropes. *arXiv e-prints*, page arXiv:2010.02619, October 2020.

[CDGS14] Sayan Chakrabarti, Térence Delsate, Norman Gürlebeck, and Jan Steinhoff. I-Q Relation for Rapidly Rotating Neutron Stars. *Phys. Rev. Lett.*, 112(20):201102, May 2014.

[CdK15] Cecilia Chirenti, Gibran H. de Souza, and Wolfgang Kastaun. Fundamental oscillation modes of neutron stars: Validity of universal relations. *Phys. Rev. D*, 91(4):044034, February 2015.

[CDK+17] Michael Coughlin, Tim Dietrich, Kyohei Kawaguchi, Stephen Smartt, Christopher Stubbs, and Maximiliano Ujevic. Toward Rapid Transient Identification and Characterization of Kilonovae. *ApJ*, 849(1):12, November 2017.

[CFR+20] H. T. Cromartie, E. Fonseca, S. M. Ransom, P. B. Demorest, Z. Arzoumanian, H. Blumer, P. R. Brook, M. E. DeCesar, T. Dolch, J. A. Ellis, R. D. Ferdman, E. C. Ferrara, N. Garver-Daniels, P. A. Gentile, M. L. Jones, M. T. Lam, D. R. Lorimer, R. S. Lynch, M. A. McLaughlin, C. Ng, D. J. Nice, T. T. Pennucci, R. Spiewak, I. H. Stairs, K. Stovall, J. K. Swiggum, and W. W. Zhu. Relativistic Shapiro delay measurements of an extremely massive millisecond pulsar. *Nature Astronomy*, 4:72–76, January 2020.

[Cha70] S. Chandrasekhar. Solutions of Two Problems in the Theory of Gravitational Radiation. *Phys. Rev. Lett.*, 24(11):611–615, March 1970.

[CSLL14] T. K. Chan, Y. H. Sham, P. T. Leung, and L. M. Lin. Multipolar universal relations between f -mode frequency and tidal deformability of compact stars. *Phys. Rev. D*, 90(12):124023, December 2014.

[DGKK13] Daniela D. Doneva, Erich Gaertig, Kostas D. Kokkotas, and Christian Krüger. Gravitational wave asteroseismology of fast rotating neutron stars with realistic equations of state. *Phys. Rev. D*, 88(4):044052, August 2013.

[DH01] F. Douchin and P. Haensel. A unified equation of state of dense matter and neutron star structure. *A&A*, 380:151–167, December 2001.

[DK15] Daniela D. Doneva and Kostas D. Kokkotas. Asteroseismology of rapidly rotating neutron stars: An alternative approach. *Phys. Rev. D*, 92(12):124004, December 2015.

[DN09] Thibault Damour and Alessandro Nagar. Relativistic tidal properties of neutron stars. *Phys. Rev. D*, 80(8):084035, October 2009.

[DRB+18] Tim Dietrich, David Radice, Sebastiano Bernuzzi, Francesco Zappa, Albino Perego, Bernd Brügmann, Swami Vivekanandji Chaurasia, Reetika Dudi, Wolfgang Tichy, and Maximiliano Ujevic. CoRe database of binary neutron star merger waveforms. *Classical and Quantum Gravity*, 35(24):24LT01, December 2018.

[DYSK14] Daniela D. Doneva, Stoytcho S. Yazadjiev, Nikolaos Stergioulas, and Kostas D. Kokkotas. Breakdown of I-Love-Q Universality in Rapidly Rotating Relativistic Stars. *ApJ*, 781(1):L6, January 2014.

[Fav14] Marc Favata. Systematic Parameter Errors in Inspiraling Neutron Star Binaries. *Phys. Rev. Lett.*, 112(10):101101, March 2014.

[FH08] Éanna É. Flanagan and Tanja Hinderer. Constraining neutron-star tidal Love numbers with gravitational-wave detectors. *Phys. Rev. D*, 77(2):021502, January 2008.

[FIS88] John L. Friedman, James R. Ipser, and Rafael D. Sorkin. Turning Point Method for Axisymmetric Stability of Rotating Relativistic Stars. *ApJ*, 325:722, February 1988.

[FS78] J. L. Friedman and B. F. Schutz. Lagrangian perturbation theory of nonrelativistic fluids. *ApJ*, 221:937–957, May 1978.

[GAC+20] He Gao, Shun-Ke Ai, Zhou-Jian Cao, Bing Zhang, Zhen-Yu Zhu, Ang Li, Nai-Bo Zhang, and Andreas Bauswein. Relation between gravitational mass and baryonic mass for non-rotating and rapidly rotating neutron stars. *Frontiers of Physics*, 15(2):24603, January 2020.

[Ger70] Robert Geroch. Multipole Moments. II. Curved Space. *Journal of Mathematical Physics*, 11(8):2580–2588, August 1970.

[GHL+20] S. K. Greif, K. Hebeler, J. M. Lattimer, C. J. Pethick, and A. Schwenk. Equation of State Constraints from Nuclear Physics, Neutron Star Masses, and Future Moment of Inertia Measurements. *ApJ*, 901(2):155, October 2020.

[GK11] Erich Gaertig and Kostas D. Kokkotas. Gravitational wave asteroseismology with fast rotating neutron stars. *Phys. Rev. D*, 83(6):064031, March 2011.

[HCPR14] B. Haskell, R. Ciolfi, F. Pannarale, and L. Rezzolla. On the universality of I-Love-Q relations in magnetized neutron stars. *MNRAS*, 438(1):L71–L75, February 2014.

[Hin08] Tanja Hinderer. Tidal Love Numbers of Neutron Stars. *ApJ*, 677(2):1216–1220, April 2008.

[Hin09] Tanja Hinderer. Erratum: "Tidal Love Numbers of Neutron Stars". *ApJ*, 697(1):964, May 2009.

[HKK+13] Kenta Hotokezaka, Kenta Kiuchi, Koutarou Kyutoku, Hirotada Okawa, Yuichiro Sekiguchi, Masaru Shibata, and Keisuke Taniguchi. Mass ejection from the merger of binary neutron stars. *Phys. Rev. D*, 87(2):024001, January 2013.

[HL99] Wynn C. G. Ho and Dong Lai. Resonant tidal excitations of rotating neutron stars in coalescing binaries. *MNRAS*, 308(1):153–166, September 1999.

[HS19] Sophia Han and Andrew W. Steiner. Tidal deformability with sharp phase transitions in binary neutron stars. *Phys. Rev. D*, 99(8):083014, April 2019.

[KAA01] K. D. Kokkotas, T. A. Apostolatos, and N. Andersson. The inverse problem for pulsating neutron stars: a 'fingerprint analysis' for the supranuclear equation of state. *MNRAS*, 320(3):307–315, January 2001.

[KGK10] Christian Krüger, Erich Gaertig, and Kostas D. Kokkotas. Oscillations and instabilities of fast and differentially rotating relativistic stars. *Phys. Rev. D*, 81(8):084019, April 2010.

[KK20a] Christian J. Krüger and Kostas D. Kokkotas. Dynamics of fast rotating neutron stars: An approach in the Hilbert gauge. *Phys. Rev. D*, 102(6):064026, September 2020.

[KK20b] Christian J. Krüger and Kostas D. Kokkotas. Fast Rotating Relativistic Stars: Spectra and Stability without Approximation. *Phys. Rev. Lett.*, 125(11):111106, September 2020.

[KKK+20] Kenta Kiuchi, Kyohei Kawaguchi, Koutarou Kyutoku, Yuichiro Sekiguchi, and Masaru Shibata. Sub-radian-accuracy gravitational waves from coalescing binary neutron stars in numerical relativity. II. Systematic study on the equation of state, binary mass, and mass ratio. *Phys. Rev. D*, 101(8):084006, April 2020.

[KS92] K. D. Kokkotas and B. F. Schutz. W-modes—A new family of normal modes of pulsating relativistic stars. *MNRAS*, 255:119–128, March 1992.

[KS99] Kostas D. Kokkotas and Bernd G. Schmidt. Quasi-Normal Modes of Stars and Black Holes. *Living Reviews in Relativity*, 2(1):2, September 1999.

[Lan14] Philippe Landry. Tidal Deformations of Compact Bodies in General Relativity. Master's thesis, The University of Guelph, Ontario, Canada, 2014.

[LK12] D. R. Lorimer and M. Kramer. *Handbook of Pulsar Astronomy*. 2012.

[LLL10] H. K. Lau, P. T. Leung, and L. M. Lin. Inferring Physical Parameters of Compact Stars from their f-mode Gravitational Wave Signals. *ApJ*, 714(2):1234–1238, May 2010.

[LNO06] Benjamin D. Lackey, Mohit Nayyar, and Benjamin J. Owen. Observational constraints on hyperons in neutron stars. *Phys. Rev. D*, 73(2):024021, January 2006.

[Lov09] A. E. H. Love. The Yielding of the Earth to Disturbing Forces. *Proceedings of the Royal Society of London Series A*, 82(551):73–88, February 1909.

[LP01] J. M. Lattimer and M. Prakash. Neutron Star Structure and the Equation of State. *ApJ*, 550(1):426–442, March 2001.

[LS05] James M. Lattimer and Bernard F. Schutz. Constraining the Equation of State with Moment of Inertia Measurements. *ApJ*, 629(2):979–984, August 2005.

[MCF+13] Andrea Maselli, Vitor Cardoso, Valeria Ferrari, Leonardo Gualtieri, and Paolo Pani. Equation-of-state-independent relations in neutron stars. *Phys. Rev. D*, 88(2):023007, July 2013.

[MMGF14] Grégoire Martinon, Andrea Maselli, Leonardo Gualtieri, and Valeria Ferrari. Rotating protoneutron stars: Spin evolution, maximum mass, and I-Love-Q relations. *Phys. Rev. D*, 90(6):064026, September 2014.

[MOHN17] Miguel Marques, Micaela Oertel, Matthias Hempel, and Jérôme Novak. New temperature dependent hyperonic equation of state: Application to rotating neutron star models and I -Q relations. *Phys. Rev. C*, 96(4):045806, October 2017.

[MS96] Horst Müller and Brian D. Serot. Relativistic Mean-Field Theory and the High-Density Nuclear Equation of State. In *APS Meeting Abstracts*, APS Meeting Abstracts, page E7.04, May 1996.

[MWRS18] Elias R. Most, Lukas R. Weih, Luciano Rezzolla, and Jürgen Schaffner-Bielich. New Constraints on Radii and Tidal Deformabilities of Neutron Stars from GW170817. *Phys. Rev. Lett.*, 120(26):261103, June 2018.

[NP20] Rana Nandi and Subrata Pal. Finding quark content of neutron stars in light of GW170817. *arXiv e-prints*, page arXiv:2008.10943, August 2020.

[NSGE98] T. Nozawa, N. Stergioulas, E. Gourgoulhon, and Y. Eriguchi. Construction of highly accurate models of rotating neutron stars - comparison of three different numerical schemes. *A&AS*, 132:431–454, November 1998.

[PA14] George Pappas and Theocharis A. Apostolatos. Effectively Universal Behavior of Rotating Neutron Stars in General Relativity Makes Them Even Simpler than Their Newtonian Counterparts. *Phys. Rev. Lett.*, 112(12):121101, March 2014.

[PGMF18] Paolo Pani, Leonardo Gualtieri, Andrea Maselli, and Valeria Ferrari. Recent progress on the tidal deformability of spinning compact objects. In Massimo Bianchi, Robert T. Jansen, and Remo Ruffini, editors, *Fourteenth Marcel Grossmann Meeting—MG14*, pages 1587–1593, January 2018.

[PS17] Vasileios Paschalidis and Nikolaos Stergioulas. Rotating stars in relativity. *Living Reviews in Relativity*, 20(1):7, November 2017.

[RBC+13] Jocelyn S. Read, Luca Baiotti, Jolien D. E. Creighton, John L. Friedman, Bruno Giacomazzo, Koutarou Kyutoku, Charalampos Markakis, Luciano Rezzolla, Masaru Shibata, and Keisuke Taniguchi. Matter effects on binary neutron star waveforms. *Phys. Rev. D*, 88(4):044042, August 2013.

[RFR+14] Carl L. Rodriguez, Benjamin Farr, Vivien Raymond, Will M. Farr, Tyson B. Littenberg, Diego Fazi, and Vicky Kalogera. Basic Parameter Estimation of Binary Neutron Star Systems by the Advanced LIGO/Virgo Network. *ApJ*, 784(2):119, April 2014.

[RLOF09] Jocelyn S. Read, Benjamin D. Lackey, Benjamin J. Owen, and John L. Friedman. Constraints on a phenomenologically parametrized neutron-star equation of state. *Phys. Rev. D*, 79(12):124032, June 2009.

[RMW18] Luciano Rezzolla, Elias R. Most, and Lukas R. Weih. Using Gravitational-wave Observations and Quasi-universal Relations to Constrain the Maximum Mass of Neutron Stars. *ApJ*, 852(2):L25, January 2018.

[RÖP18] Carolyn A. Raithel, Feryal Özel, and Dimitrios Psaltis. Tidal Deformability from GW170817 as a Direct Probe of the Neutron Star Radius. *ApJ*, 857(2):L23, April 2018.

[RPBZ18] David Radice, Albino Perego, Sebastiano Bernuzzi, and Bing Zhang. Long-lived remnants from binary neutron star mergers. *MNRAS*, 481(3):3670–3682, December 2018.

[RPH+18] David Radice, Albino Perego, Kenta Hotokezaka, Steven A. Fromm, Sebastiano Bernuzzi, and Luke F. Roberts. Binary Neutron Star Mergers: Mass Ejection, Electromagnetic Counterparts, and Nucleosynthesis. *ApJ*, 869(2):130, Dec 2018.

[RT16] Luciano Rezzolla and Kentaro Takami. Gravitational-wave signal from binary neutron stars: A systematic analysis of the spectral properties. *Phys. Rev. D*, 93(12):124051, June 2016.

[RW57] Tullio Regge and John A. Wheeler. Stability of a Schwarzschild Singularity. *Physical Review*, 108(4):1063–1069, November 1957.

[SAF04] Nikolaos Stergioulas, Theocharis A. Apostolatos, and José A. Font. Non-linear pulsations in differentially rotating neutron stars: mass-shedding-induced damping and splitting of the fundamental mode. *MNRAS*, 352(4):1089–1101, August 2004.

[SBZJ11] Nikolaos Stergioulas, Andreas Bauswein, Kimon Zagkouris, and Hans-Thomas Janka. Gravitational waves and non-axisymmetric oscillation modes in mergers of compact object binaries. *MNRAS*, 418(1):427–436, November 2011.

[Sch08] B. F. Schutz. Asteroseismology of neutron stars and black holes. In *Journal of Physics Conference Series*, volume 118 of *Journal of Physics Conference Series*, page 012005, October 2008.

[SF95] Nikolaos Stergioulas and John L. Friedman. Comparing Models of Rapidly Rotating Relativistic Stars Constructed by Two Numerical Methods. *ApJ*, 444:306, May 1995.

[SKK+16] Yuichiro Sekiguchi, Kenta Kiuchi, Koutarou Kyutoku, Masaru Shibata, and Keisuke Taniguchi. Dynamical mass ejection from the merger of asymmetric binary neutron stars: Radiation-hydrodynamics study in general relativity. *Phys. Rev. D*, 93(12):124046, June 2016.

[SLB16] A. W. Steiner, J. M. Lattimer, and E. F. Brown. Neutron star radii, universal relations, and the role of prior distributions. *European Physical Journal A*, 52:18, February 2016.

[TC67] Kip S. Thorne and Alfonso Campolattaro. Non-Radial Pulsation of General-Relativistic Stellar Models. I. Analytic Analysis for L >= 2. *ApJ*, 149:591, September 1967.

[TH85] Kip S. Thorne and James B. Hartle. Laws of motion and precession for black holes and other bodies. *Phys. Rev. D*, 31(8):1815–1837, April 1985.

[Tho98] Kip S. Thorne. Tidal stabilization of rigidly rotating, fully relativistic neutron stars. *Phys. Rev. D*, 58(12):124031, December 1998.

[TL05a] L. K. Tsui and P. T. Leung. Probing the Interior of Neutron Stars with Gravitational Waves. *Phys. Rev. Lett.*, 95(15):151101, October 2005.

[TL05b] L. K. Tsui and P. T. Leung. Universality in quasi-normal modes of neutron stars. *MNRAS*, 357(3):1029–1037, March 2005.

[VK19] Sebastian H. Völkel and Kostas D. Kokkotas. On the inverse spectrum problem of neutron stars. *Classical and Quantum Gravity*, 36(11):115002, June 2019.

[VSB20] Stamatis Vretinaris, Nikolaos Stergioulas, and Andreas Bauswein. Empirical relations for gravitational-wave asteroseismology of binary neutron star mergers. *Phys. Rev. D*, 101(8):084039, April 2020.

[WFF88] R. B. Wiringa, V. Fiks, and A. Fabrocini. Equation of state for dense nucleon matter. *Phys. Rev. C*, 38(2):1010–1037, August 1988.

[WLCZ19] De-Hua Wen, Bao-An Li, Hou-Yuan Chen, and Nai-Bo Zhang. Relations among the characteristics of f-mode oscillations, tidal deformability and radii of canonical neutron stars. In *Xiamen-CUSTIPEN Workshop on the Equation of State of Dense Neutron-Rich Matter in the Era of Gravitational Wave Astronomy*, volume 2127 of *American Institute of Physics Conference Series*, page 020034, July 2019.

[YRKE02] Shin'ichirou Yoshida, Luciano Rezzolla, Shigeyuki Karino, and Yoshiharu Eriguchi. Frequencies of f-Modes in Differentially Rotating Relativistic Stars and Secular Stability Limits. *ApJ*, 568(1):L41–L44, March 2002.

[YTK20a] Shotaro Yamasaki, Tomonori Totani, and Kenta Kiuchi. FRB 181112 as a Rapidly-Rotating Massive Neutron Star just after a Binary Neutron Star Merger?: Implications for Future Constraints on Neutron Star Equations of State. *arXiv e-prints*, page arXiv:2010.07796, October 2020.

[YTK20b] Shotaro Yamasaki, Tomonori Totani, and Kenta Kiuchi. FRB 181112 as a Rapidly-Rotating Massive Neutron Star just after a Binary Neutron Star Merger?: Implications for Future Constraints on Neutron Star Equations of State. *arXiv e-prints*, page arXiv:2010.07796, October 2020.

[YY13] Kent Yagi and Nicolás Yunes. I-Love-Q: Unexpected Universal Relations for Neutron Stars and Quark Stars. *Science*, 341(6144):365–368, July 2013.

[YY16] Kent Yagi and Nicolás Yunes. Binary Love relations. *Classical and Quantum Gravity*, 33(13):13LT01, July 2016.

[YY17] Kent Yagi and Nicolás Yunes. Approximate universal relations for neutron stars and quark stars. *Phys. Rep.*, 681:1–72, April 2017.

[ZBR+18] Francesco Zappa, Sebastiano Bernuzzi, David Radice, Albino Perego, and Tim Dietrich. Gravitational-Wave Luminosity of Binary Neutron Stars Mergers. *Phys. Rev. Lett.*, 120(11):111101, March 2018.

[ZL18] Tianqi Zhao and James M. Lattimer. Tidal deformabilities and neutron star mergers. *Phys. Rev. D*, 98(6):063020, September 2018.

Printed in the United States
by Baker & Taylor Publisher Services